INTEGRATED WASTE MANAGEMENT APPROACHES FOR FOOD AND AGRICULTURAL BYPRODUCTS

INTEGRATED WASTE MANAGEMENT APPROACHES FOR FOOD AND AGRICULTURAL BYPRODUCTS

Edited by
Tawheed Amin, PhD
Omar Bashir
Shakeel Ahmad Bhat
Muneeb Ahmad Malik

First edition published 2023

Apple Academic Press Inc.
1265 Goldenrod Circle, NE,
Palm Bay, FL 32905 USA
760 Laurentian Drive, Unit 19,
Burlington, ON L7N 0A4, CANADA

CRC Press
6000 Broken Sound Parkway NW,
Suite 300, Boca Raton, FL 33487-2742 USA
4 Park Square, Milton Park,
Abingdon, Oxon, OX14 4RN UK

© 2023 by Apple Academic Press, Inc.

Apple Academic Press exclusively co-publishes with CRC Press, an imprint of Taylor & Francis Group, LLC

Reasonable efforts have been made to publish reliable data and information, but the authors, editors, and publisher cannot assume responsibility for the validity of all materials or the consequences of their use. The authors, editors, and publishers have attempted to trace the copyright holders of all material reproduced in this publication and apologize to copyright holders if permission to publish in this form has not been obtained. If any copyright material has not been acknowledged, please write and let us know so we may rectify in any future reprint.

Except as permitted under U.S. Copyright Law, no part of this book may be reprinted, reproduced, transmitted, or utilized in any form by any electronic, mechanical, or other means, now known or hereafter invented, including photocopying, microfilming, and recording, or in any information storage or retrieval system, without written permission from the publishers.

For permission to photocopy or use material electronically from this work, access www.copyright.com or contact the Copyright Clearance Center, Inc. (CCC), 222 Rosewood Drive, Danvers, MA 01923, 978-750-8400. For works that are not available on CCC please contact mpkbookspermissions@tandf.co.uk

Trademark notice: Product or corporate names may be trademarks or registered trademarks and are used only for identification and explanation without intent to infringe.

Library and Archives Canada Cataloguing in Publication

Title: Integrated waste management approaches for food and agricultural byproducts / edited by Tawheed Amin, PhD, Omar Bashir, Shakeel Ahmad Bhat, Muneeb Ahmad Malik.
Names: Amin, Tawheed, editor. | Bashir, Omar, editor. | Bhat, Shakeel Ahmad, editor. | Malik, Muneeb Ahmad, editor.
Description: First edition. | Includes bibliographical references and index.
Identifiers: Canadiana (print) 20220449635 | Canadiana (ebook) 20220449759 | ISBN 9781774910160 (hardcover) | ISBN 9781774910177 (softcover) | ISBN 9781003282327 (ebook)
Subjects: LCSH: Food waste. | LCSH: Food waste—Prevention. | LCSH: Food waste—Environmental aspects. | LCSH: Agricultural wastes. | LCSH: Agricultural wastes—Recycling. | LCSH: Agricultural wastes—Environmental aspects.
Classification: LCC TD804 .I58 2023 | DDC 363.72/88—dc23

Library of Congress Cataloging-in-Publication Data

··

CIP data on file with US Library of Congress

··

ISBN: 978-1-77491-016-0 (hbk)
ISBN: 978-1-77491-017-7 (pbk)
ISBN: 978-1-00328-232-7 (ebk)

About the Editors

Tawheed Amin, PhD
*Assistant Professor-cum-Junior Scientist
(Food Science and Technology),
Division of Food Science and Technology of
Kashmir, Sher-e-Kashmir University of
Agricultural Sciences and Technology,
Jammu and Kashmir, India*

Tawheed Amin, PhD, is presently working as an Assistant Professor-cum-Junior Scientist (Food Science and Technology) in the Division of Food Science and Technology of Kashmir (SKUAST-Kashmir). He has been a Food Safety Officer in the Drug and Food Control Organization, Government of Jammu & Kashmir. He did his PhD in Food Technology at the Division of Food Science and Technology, Sher-e-Kashmir University of Agricultural Sciences and Technology-Kashmir (SKUAST-K), Shalimar, Srinagar, Jammu and Kashmir, India. He did his MTech in Food Technology from Amity University, Noida, Uttar Pradesh, and BTech in Food Technology from the Department of Food Technology, Islamic University of Science and Technology, Awantipora, Jammu & Kashmir, India. He is the recipient of two gold medals for being the topper in BTech (Food Technology) and PhD (Food Technology). He has also received the First Rank Holder certificate in MTech Food Technology. He has attended several national and international conferences, symposiums, and workshops and presented posters. He has to his credit several national and international papers and popular articles.

Omar Bashir
Assistant Professor-cum-Junior Scientist (Food Technology and Nutrition), Department of Food Technology and Nutrition, Lovely Professional University, Punjab, India.

Omar Bashir, PhD, is presently working as an Assistant Professor-cum-Junior Scientist (Food Technology and Nutrition) in the Department of Food Technology and Nutrition, Lovely Professional University, Punjab, India. He did his PhD in Food Technology at the Division of Food Science and Technology at Sher-e-Kashmir University of Agricultural Sciences and Technology, Srinagar, Jammu and Kashmir, India. He has published more than 30 original research and review articles in peer- reviewed high-impact journals along with many national and international book chapters and magazine articles. He has attended several national and international conferences and workshops. He has been awarded by Ministry of Food Processing Industries (MOFPI), India, for contributing to the concept of emerging technologies for powdered foods. He also has to his credit editorial board membership for various reputed journals.

Shakeel Ahmad Bhat
PhD Scholar, College of Agricultural Engineering, Sher-e-Kashmir University of Agricultural Sciences and Technology (SKUAST-K), Jammu and Kashmir, India

Shakeel Ahmad Bhat is a PhD Scholar in Agricultural Engineering (Soil and Water Engineering) at the College of Agricultural Engineering at Sher-e-Kashmir University of Agricultural Sciences and Technology (SKUAST-K). He is currently working on hydroponics technology. He is the author of more than 15 scientific articles and five book chapters. He has presented and participated in numerous state, national, and international conferences, seminars, and workshops. He is also a reviewer for various international journals and holds life memberships in various international organizations.

About the Editors

Muneeb Ahmad Malik
Senior Research Fellow, Indian Council of Medical Research (ICMR), India

Muneeb Ahmad Malik holds bachelor's and master's degrees in Food Technology. He is presently a Senior Research Fellow with the Indian Council of Medical Research (ICMR) and is pursuing a PhD in the Department of Food Technology, Jamia Hamdard University, New Delhi, India. He has published several peer-reviewed articles in the area of probiotics and vitamin D deficiency. His current research area involves studies related to vitamin D deficiency and exploration of vegan sources of vitamin D.

Contents

Contributors .... *xi*

Abbreviations .... *xv*

Foreword .... *xix*

Preface .... *xxi*

1. **Characterization of Food and Agricultural Wastes: Global Scenario of Waste Generation** 1

   Entesar Hanan, Farhan J. Ahmad, Vasudha Sharma, Omar Bashir, Muneeb Malik, and Yasmeena Jan

2. **Food Processing Byproducts: Their Applications as Sources of Valuable Bioenergy and Recoverable Products** 25

   Ufaq Fayaz, Iqra Bashir, Jibreez Fayaz, Omar Bashir, Sobiya Manzoor, Tawheed Amin, and Shakeel Ahmad Bhat

3. **Waste to Wealth: Reduction, Reuse, and Recycling of Food and Agricultural Waste** 81

   Rubiya Rashid, F. A. Masoodi, Sajad Mohd Wani, Shakeel Ahmad Bhat, Shaziya Manzoor, Omar Bashir, Rouf Ahmad Bhat, and Ab. Waheed Wani

4. **Basic and Modern Environmental Management Practices for Food and Agricultural Waste Management** 113

   Quraazah Akeemu Amin, Towseef Ahmad Wani, Tawheed Amin, Afsah Iqbal Nahvi, Taha Mukhtar, Nazrana Rafique, and Shubli Bashir

5. **Food Waste Management: Approaches to Achieve Food Security and Sustainability** 141

   Afnan Ashraf, Syed Anam Ul Haq, Shabir Hassan, Shakeel Ahmad Bhat, Shahid Qayoom, Abbas Ahmad Mir, and Mahsa Mirzakhani

6. **Impact of Food and Agricultural Wastes on the Environment: Management Strategies and Regulations to Curb Wastes** 165

   Yasmeena Jan, Muneeb Malik, Afrozul Haq, Bibhu Prasad Panda, Mifftha Yaseen, Entesar Hanan, Ishfaq Yaseen, and Asif Rafiq

7. **Challenges and Opportunities Associated with Food and Agricultural Waste Management Across the Globe**.............. 201

Drishti Kadian, Syed Mansha Rafiq, Nikunj Sharma, Syed Insha Rafiq, Syed Anam Ul Haq, R. M. Shukla, Faizan Masoudi, and Insha Nazir

Index... 219

Contributors

Farhan J. Ahmad
Department of Pharmaceutics, School of Pharmaceutical Education and Research, Jamia Hamdard, New Delhi – 110062, India

Quraazah Akeemu Amin
Division of Food Science and Technology, Sher-e-Kashmir University of Agricultural Sciences and Technology of Kashmir, Jammu and Kashmir – 190025, India, E-mail: widaad57@gmail.com

Tawheed Amin
Division of Food Science and Technology, Sher-e-Kashmir University of Agricultural Sciences and Technology of Kashmir, Jammu and Kashmir – 190025, India

Afnan Ashraf
Department of Soil and Water Engineering, Punjab Agricultural University, Ludhiana, Punjab, India

Iqra Bashir
Division of Food Science and Technology, Sher-e-Kashmir University of Agricultural Sciences and Technology of Kashmir, Shalimar – 190025, Jammu and Kashmir, India

Omar Bashir
Department of Food Technology and Nutrition, Lovely Professional University, Punjab, India

Shubli Bashir
Division of Food Science and Technology, Sher-e-Kashmir University of Agricultural Sciences and Technology of Kashmir, Jammu and Kashmir – 190025, India

Rouf Ahmad Bhat
Division of Environmental Science, Sher-e-Kashmir University of Agricultural Sciences and Technology of Kashmir, Jammu and Kashmir – 190025, India

Shakeel Ahmad Bhat
College of Agricultural Engineering and Technology, Sher-e-Kashmir University of Agricultural Sciences and Technology of Kashmir, Shalimar, Jammu and Kashmir, India, E-mail: wakeelbhat@gmaill.com

Jibreez Fayaz
Department of Food Technology, Islamic University of Science and Technology, Awantipora – 192122, Jammu and Kashmir, India

Ufaq Fayaz
Division of Food Science and Technology, Sher-e-Kashmir University of Agricultural Sciences and Technology of Kashmir, Shalimar – 190025, Jammu and Kashmir, India

Gousia Gani
Division of Food Science and Technology, Sher-e-Kashmir University of Agricultural Sciences and Technology of Kashmir, Jammu and Kashmir – 190025, India

Entesar Hanan
Department of Food Technology, School of Interdisciplinary Sciences and Technology, Jamia Hamdard, New Delhi – 110062, India, E-mail: entesarhanan@yahoo.in

Afrozul Haq
Department of Food Technology, School of Interdisciplinary Sciences and Technology, Jamia Hamdard, New Delhi, India

Syed Anam Ul Haq
Department of Plant Biotechnology, Sher-e-Kashmir University of Agricultural Sciences and Technology, Shalimar, Srinagar, Jammu and Kashmir, India

Shabir Hassan
Department of Medicine, Harvard Medical School, Boston, United States

Nusrat Jan
Division of Food Science and Technology, Sher-e-Kashmir University of Agricultural Sciences and Technology of Kashmir, Shalimar – 190025, Jammu and Kashmir, India

Yasmeena Jan
Department of Food Technology, School of Interdisciplinary Sciences and Technology, Jamia Hamdard, New Delhi – 110062, India, E-mail: yasmeenajan070@gmail.com

Drishti Kadian
Department of Dairy Technology, ICAR-National Dairy Research Institute, Karnal – 132001, Haryana, India

Muneeb Malik
Department of Food Technology, School of Interdisciplinary Sciences and Technology, Jamia Hamdard, New Delhi – 110062, India

Shaziya Manzoor
Department of Food Science and Technology, University of Kashmir, Srinagar, Jammu and Kashmir, India

Sobiya Manzoor
Department of Food Technology, Islamic University of Science and Technology, Awantipora – 192122, Jammu and Kashmir, India

F. A. Masoodi
Department of Food Science and Technology, University of Kashmir, Srinagar, Jammu and Kashmir, India

Faizan Masuadi
College of Agricultural Engineering, Sher-e-Kashmir University of Agricultural Sciences and Technology, Shalimar, Srinagar, Jammu and Kashmir, India

Abbas Ahmad Mir
Department of Geography and Disaster Management, University of Kashmir, Jammu and Kashmir, India

Mahsa Mirzakhani
Department of Natural Resources, Isfahan, University of Technology, Iran

Taha Mukhtar
Division of Food Science and Technology, Sher-e-Kashmir University of Agricultural Sciences and Technology of Kashmir, Shalimar – 190025, Jammu and Kashmir, India

Contributors

Afsah Iqbal Nahvi
Division of Food Science and Technology, Sher-e-Kashmir University of Agricultural Sciences and Technology of Kashmir, Jammu and Kashmir – 190025, India

Insha Nazir
Department of Floriculture and Landscaping, Sher-e-Kashmir University of Agricultural Sciences and Technology, Shalimar, Srinagar, Jammu and Kashmir, India

Bibhu Prasad Panda
Department of Pharmacognosy and Phytochemistry, School of Pharmaceutical Education and Research, Jamia Hamdard, New Delhi, India

Shahid Qayoom
Department of Fruit Science, Sher-e-Kashmir University of Agricultural Sciences and Technology, Shalimar, Srinagar, Jammu and Kashmir, India

Asif Rafiq
College of Temperate Sericulture, Sher-e-Kashmir University of Agricultural Sciences and Technology, Shalimar, Srinagar, Jammu and Kashmir, India

Syed Insha Rafiq
Department of Dairy Technology, ICAR-National Dairy Research Institute, Karnal – 132001, Haryana, India

Syed Mansha Rafiq
Department of Food Science and Technology, National Institute of Food Technology Entrepreneurship and Management, Sonipat – 131028, Haryana, India,
E-mail: mansharafiq@gmail.com

Nazrana Rafique
Division of Food Science and Technology, Sher-e-Kashmir University of Agricultural Sciences and Technology of Kashmir, Jammu and Kashmir – 190025, India

Rubiya Rashid
Department of Food Science and Technology, University of Kashmir, Srinagar, Jammu and Kashmir, India, E-mail: rubiyarashideng@gmail.com

Nikunj Sharma
Department of Food Science and Technology, National Institute of Food Technology Entrepreneurship and Management, Sonipat – 131028, Haryana, India

Vasudha Sharma
Food Technology, School of Interdisciplinary Sciences and Technology, Jamia Hamdard, New Delhi – 110062, India

R. M. Shukla
College of Agricultural Engineering, Sher-e-Kashmir University of Agricultural Sciences and Technology, Shalimar, Srinagar, Jammu and Kashmir, India

Ab. Waheed Wani
Department of Fruit Science, Sher-e-Kashmir University of Agricultural Sciences and Technology, Shalimar, Srinagar, Jammu and Kashmir, India

Sajad Mohd. Wani
Division of Food Science and Technology, Sher-e-Kashmir University of Agricultural Sciences and Technology of Kashmir, Srinagar – 190025, Jammu and Kashmir, India

Towseef Ahmad Wani
Division of Food Science and Technology, Sher-e-Kashmir University of Agricultural Sciences and Technology of Kashmir, Jammu and Kashmir – 190025, India

Ishfaq Yaseen
Department of Management, Islamia College, University of Kashmir, Jammu and Kashmir, India

Mifftha Yaseen
Department of Food Technology, School of Interdisciplinary Sciences and Technology, Jamia Hamdard, New Delhi, India

Abbreviations

AAC	aliphatic-aromatic copolyesters
AD	anaerobic digestion
AFW	agricultural and food waste
ALA	α-linolenic acid
AWMS	agricultural waste management system
BES	bioelectrochemical system
BOD	biological oxygen demand
BSE	bovine spongiform encephalopathy
C/N	carbon-nitrogen
CaC_2	calcium carbide
CH_4	methane
CIPET	Central Institute of Plastics Engineering and Technology
CO	carbon monoxide
CO_2	carbon dioxide
COD	chemical oxygen demand
CPCB	central pollution control board
CS_2	carbon disulfide
DHA	docosahexaenoic acid
DME	dimethyl ether
EFSA	European Food Safety Authority
EU	European Union
FAO	Food and Agriculture Organization
FFA	free fatty acids
FICCI	Federation of Indian Chambers of Commerce and Industry
FL	food loss
FSC	food supply chain
FVW	fruit and vegetable waste
FWH	FW hydrolysate
GA	glucoamylase
GHG	greenhouse gases

GRAS	generally recognized as safe
H2	hydrogen
HDPE	high-density polyethylene
HTC	hydrothermal carbonization
HTL	hydrothermal liquefaction
IPCC	intergovernmental panel on climate change
LDPE	low-density polyethylene
MASE	microwave-assisted solvent extraction
MMT	million metric tons
MNRE	ministry of new and renewable energy
MT	million tons
NaCn	sodium cyanide
NEERI	National Engineering and Environmental Research Institute
NH_3	ammonia
NPK	nitrogen-phosphorus-potassium
NPMCR	national policy for management of crop residues
OECD	Organization for Economic Cooperation and Development
PBAT	polybutylene adipate/terephthalate
PBS	polybutylene succinate
PBSA	polybutylene succinate adipate
PCL	poly ε-caprolactone
PET	polyethylene terephthalate
PGA	polyglycolic acid
PHA	polyhydroxyalkanoates
PHB	polyhydroxy butyrate
PLA	polylactic acid
PM	particulate matter
POME	palm oil mill effluent
PS	polystyrene
PTMAT	poly methylene adipate/terephthalate
PUFAs	poly-unsaturated fatty acids
RFID	radio frequency identification detectors
SCACA	short-chain aliphatic carboxylic acid
SCP	single-cell protein
SDG	sustainability development goal

SFE	supercritical fluid extraction
SiC	silicon carbide
SMF	submerged fermentation
SSF	solid-state fermentation
TTI	time-temperature indicators
UAE	ultrasound-assisted extraction
UNFCC	united nations framework convention on climate change
USEPA	United State Environmental Protection Agency
UV	ultraviolet
VFA	volatile fatty acid
VOC	volatile organic compound
WM	waste management
WRAP	water resource action program
WRRF	water resource recovery facility

Foreword

I am delighted to write the foreword for the book *Integrated Waste Management Approaches for Food and Agricultural Byproducts*. This book presents new methods and technologies to combat the problems associated with food waste management and focuses on the importance of integrated waste management approaches in minimizing the ill effects on the environment. The technologies related to waste treatment are economical, eco-friendly, and bring economic returns, and can be applied to most of the developing countries where waste treatment technologies are composting, anaerobic digestion, recycling of plastic, and agricultural waste in construction. The natural resource base in our world today is exposed to constantly increasing pressures. Environmental problems are on the increase in developing countries as well as in developed countries. The planning of the management and recycling of waste for all types of waste is an enormous task, which involves both logistical planning and scientific knowledge and understanding in order to balance the impact on the environment and cost-effectiveness of the process. The integrated waste management strategy relies on handling waste in a four-pronged approach: waste minimization, recycling (including composting) and energy recovery, and finally, as a last resort, landfill. Integrated waste management is comprehensive waste prevention, recycling, composting, and disposal program. An effective system considers how to prevent, recycle, and manage solid waste in ways that most effectively protect human health and the environment.

It is my hope and expectation that this book will be of interest to academicians, teachers, and researchers related to food waste management in the leading academic and research organizations globally. This book will be of prodigious value to upcoming researchers, scholars, scientists, and professionals in the environmental science and engineering fields, and global and local authorities and policymakers responsible for the management of food wastes. This book can also be a great source for designing and operating waste reuse and recycling programs.

—Prof. H. R. Naik
Dean Research, IUST, Awantipora,
Jammu and Kashmir, India

Preface

Wastage of food has become a major problem in the present times. The generation of an extensive amount of waste has created a serious threat in the world in past years that have adversely affected developing and developed countries equally. Up to one-third of all food produced for human consumption goes to waste and results in major socio-economic and environmental consequences. The wastage of food results in wasting of time as well as the energy that is used to produce it. Furthermore, waste has significant financial implications in addition to having an effect on the environment. It is quite critical to reduce the high environmental, social, and economic impacts associated with this form of waste in an increasingly resource-constrained world.

The Environmental Protection Agency (EPA) of the United States describes an integrated system for waste management (ISWM) as a complete waste decrease, compilation, composting, recycling, and discarding system. To shield human wellbeing and the normal surroundings, a capable integrated system for waste management framework is needed to lessen, reuse, recycle, and manage the refuse. Integrated waste supervision systems assess proficient waste gathering, moving, and efficient waste dumping jointly to decrease waste production and amplify waste recycling. In order to attain definite purposes and aims, integrated waste management has surfaced as a holistic tool to administer waste by merging and employing a variety of suitable methods, equipment, and organizational plans. The concept of integrated waste management evolved beyond the identification of waste; the administration frameworks are covered by numerous interrelated systems and purposes. The main aim is to identify a structure of orientation for designing and implementing new waste management systems and enhancement the existing systems. There is no ideal integrated waste management procedure that provides a single waste management technique that is appropriate for handling the entire waste in a protractible manner.

This book will provide an insight into the novel and integrated waste management techniques to students, researchers, and scientists across the globe.

CHAPTER 1

Characterization of Food and Agricultural Wastes: Global Scenario of Waste Generation

ENTESAR HANAN,[1] FARHAN J. AHMAD,[2] VASUDHA SHARMA,[1] OMAR BASHIR,[3] MUNEEB MALIK,[1] and YASMEENA JAN[1]

[1] *Department of Food Technology, School of Interdisciplinary Sciences and Technology, Jamia Hamdard, New Delhi – 110062, India, E-mail: entesarhanan@yahoo.in (E. Hanan)*

[2] *Department of Pharmaceutics, School of Pharmaceutical Education and Research, Jamia Hamdard, New Delhi – 110062, India*

[3] *Department of Food Technology and Nutrition, Lovely Professional University, Punjab, India*

ABSTRACT

Food wastage has detrimental ramifications on sustainability. Increasing concern about resource diminution, food, and nutrition security, environmental apprehensions, and greenhouse emissions has led to an exaggerated awareness of wasted food. Forlornly, individuals ravage more food than they consume. Nearly, one-third of the food which is intended for human utilization is wasted. Apart from the food being wasted at the household level, farm-level a lot of food is also wasted along with the food network in the processing, packing, and distribution. Latterly, food wastage has acquired a lot of attention on the international platform. The

---

Integrated Waste Management Approaches for Food and Agricultural Byproducts. Tawheed Amin, PhD, Omar Bashir, Shakeel Ahmad Bhat & Muneeb Ahmad Malik (Eds.)
© 2023 Apple Academic Press, Inc. Co-published with CRC Press (Taylor & Francis)

food and agriculture waste characterization is a way forward to help in waste minimization.

1.1 INTRODUCTION

Food is one of the prerequisites and fundamental requirements of human life. Its wastage has been recognized as the foremost challenge humanity is facing today (Mirabella et al., 2014). Roughly one-third of the food produced today is misspent (Stuart, 2009). Food wastage is a serious monetary and ecological problem (Kroyer, 1995). On the one hand, we have momentous food wastage, and on the other hand, approximately one in nine among the vast populace starves every night, and almost 21,000 people expire each day due to starvation-related reasons (FAO, 2008). Additionally, food wastage results in loss of time, effort, and the other resources (fertilizers, soil) that were invested in food production. Reports suggest the rise of the annual amount of food waste in metropolitan Asian countries will be from 278 to 416 million tons from 2005 to 2025 (Melikoglu et al., 2013). Roughly the world's agricultural area of about 28%, i.e., 1.4 billion hectares of productive land, is used yearly to generate food that is squandered. Waste generation depends on eating habits, culture, and geographical location. The edibility and inedibility of the food is a subjective matter. The food that may be considered inedible by some people appears to be edible for others, such as breadcrumbs, peel of an apple, etc. Besides, not all of the food or livestock product is edible. An inevitable amount of the inedible parts such as fruit stones, bones, rind of fruits, eggshells, etc., will always be there (Redlingshofer and Soyeux, 2012; Buzby and Hyman, 2012). A paramount quantity of food is also whacked along the food supply chain (FSC) during manufacturing, processing, distribution, etc. In the case of agricultural waste, in addition to the hazardous and toxic agricultural waste majority, waste is generated by animals from food processing operations and crops. An estimate of about 998 million tons of agricultural waste is produced yearly (Agamuthu, 2009). Almost 80% of the total solid wastes generation accounts to farm organic wastes, of which 5.27 kg/day/1,000 kg live weight (wet weight basis) amounts to manure production (Overcash, 1973). Reduction in food and agricultural waste is perceived as a means to reduce the production cost, improve food nutrition as well as safety, and help contribute towards the sustainable development goal (SDG). The global degree of food waste generation presents a confounding scenario, and it is imperative for us to realize the enormity of the problem. Decades ago, food wastage was not even recognized as a major waste of natural resources (Baiano, 2014). Nevertheless, the growing concern of hunger and environmental apprehensions shifted the focus on food wastage. Food wastage raises environmental concerns. According to FAO, every year, almost 3.3 billion tons of CO_2 emission takes place, which results in global warming and hence climate change (FAO, 2013). Food waste is a major contributor to greenhouse gas emanation, which has an impact on climate change. In view of the increasing greenhouse emission, the United Nations framework convention on climate change (UNFCC) signed the Paris Agreement in 2016, which aims to reinforce the global response to the menace of climate change by maintaining a global rise in temperature less than 2°C above pre-industrial levels and limits the additional rise in temperature to 1.5°C. Moreover, the United Nations sustainable development goal (SDG) 12, 'Ensure sustainable consumption and production patterns' ascertained in 2015, includes a precise goal for food waste diminution by 2030 and also on attenuation of food loss (FL) alongside the FSCs (GA, 2015). The reduction in food wastage will also have extensive insinuation on other Food related SDGs such as SDG 6 – sustainable water management, SDG 11 – sustainable cities and communities, SDG 7 – affordable and clean energy, SDG 9 – infrastructure, industry, and innovation, SDG 13 – climate change, SDG 14 – marine resources and SDG 15 – terrestrial ecosystems, forests, land, and biodiversity (FOOD, 2016). The current chapter deals with the extensive characterization of food and agricultural waste products together with a brief highlight on global waste generation.

1.2 FOOD WASTE

The United Nations Food and Agriculture Organization (FAO) defines food waste as the sum of wasted food produced in foodservice chains, with reference of 'food' to "edible products going to human utilization" (Gustavsson et al., 2011).

The European Union (EU) states food waste as "any food material, uncooked or fit to be eaten, which is thrown out or anticipated to be thrown out" (Ostergren, 2014).

It involves the food that is repudiated alongside the food chain and cannot be put to use. However, food waste can be composted, incinerated, treated through anaerobic digestion (AD), combusted for bio-energy production, disposed to sewer, spread to land, sent to landfill, dumped in open dumps, or discarded to sea (Abdel et al., 2018).

1.2.1 *VARIOUS IMPACTS OF FOOD WASTAGE*

Food wastage leads to the profligacy of the resources needed to make food that is not consumed. Apart from the wastefulness of resources such as energy, carbon, water, and nutrients, it results in manifold adverse effects. The greenhouse gases (GHGs) emanation, i.e., carbon dioxide, methane, and nitrous oxide (CO_2, CH_4, N_2O) emitted throughout the food life cycle stages, result in atmospheric heating and climate change. Water for irrigation required for the production of food results in the exhaustion of groundwater aquifers and surface water bodies, thus resulting in an adverse impact on the surface water as well as groundwater bodies. The dumping of food in landfills and junkyards can lead to various diseases (parasitic and gastrointestinal) in the population living nearby. The dumpsites further attract the carriers (flies, birds, etc.) of a communicable disease which thereby gain entry into the food chain and thus elevate the health risk. The loss of biodiversity due to the conversion of land from forests to agriculture and the widespread ecological damage for food production is yet another global-scale impact of food waste (World Biogas Association, 2018). The food waste has resulted in the countries coming together to offer promising international commitments, which include the "Paris Agreement," UNFCC, perpetrating to a combined action towards a low carbon generation by restraining global warming to 1.5–2°C above pre-industrial levels by 2100 (Sands, 1992) and the United Nation Sustainable Development Goal (SDG) 12- "Responsible consumption and production" which ensures an explicit objective of food waste reduction by 2030. The SDG 12 aims to reduce the per capita global retail and consumer food waste halve along with the reduction in food losses in the production, postharvest, and supply chain (Assembly, 2015) (Figure 1.1).

FIGURE 1.1 Impact of food waste.

1.2.2 CHARACTERIZATION OF FOOD WASTE

Different studies have classified food waste as per their own characterizations. Several researchers classified food waste on the basis of food types such as meat, fish, cereals, drinks, fruits, etc. This categorization was useful in quantifying the food waste on the basis of the mass, energy content of the waste, and economic cost (Flores et al., 1999; Malamis et al., 2015).

The cataloging of the food waste can also be done into the following categories:

1. **Nutrient Composition**: such as carbohydrate content, amount of fat (Russ et al., 2004).

2. **Chemical Constituents:** such as Carbon, Hydrogen, Nitrogen content (Roberts, 2015).

3. **Temperature of Storage:** such as frozen, ambient (Mena et al., 2011).

The United Kingdom (UK) Water Resource Action Program (WRAP) classified food waste in the following categories (Bridgwater, 2013):

1. **Stage of Supply Chain:** It includes the stage where the waste is produced, such as the harvesting stage, manufacturer stage, etc.

2. **Edibility of Waste:** On the basis of edibility, WRAP classified waste as avoidable, unavoidable, and possibly avoidable:

   i. **Avoidable Waste:** It refers to the food waste that was actually edible.

   ii. **Unavoidable Waste:** It includes the inedible parts such as bones, peel of the fruit, etc.

   iii. **Possibly Avoidable:** This category includes the food of preference which some like to eat while others discard, such as bread crusts.

Other researchers sorted food waste on the household level (Ventour, 2008; Quested and Murphy, 2014):

- Cooked and uncooked.
- Packaged and unpackaged.
  The supplementary classification of packaged food waste was done as:
  - Opened packaging;
  - Unopened packaging.
- Leftovers and untouched (Schott and Andersson, 2015).

In another study, a thorough food waste classification was done. However, the edibility and the processing of the waste were not considered in this classification (Lin et al., 2013). The waste was categorized as:

1. **Organic Crop Residue**: It includes fruit and vegetable waste.
2. **Catering Waste:** Waste from hotels, cafes, etc.
3. **Animal By-product**: Such as skin, hooves, etc.
4. **Packaging:** Packed and unpacked waste.
5. **Mixed Food Waste:** Contains more than one ingredient.
6. **Domestic Waste:** Waste from kitchens.

Another classification of fresh food waste was done on the basis of source of waste generation and is categorized into seven categories like transport, inventory, motion, waiting, overproduction, over-processing, and defects (Chabada et al., 2012).

All the above-mentioned food waste classifications lacked a standardized approach to classification. In order to grade food waste in a more appropriate manner, a more holistic approach was chosen, which included the three mainstays of sustainability are economic corollary, environmental ramifications, and social contemplation. Based on this approach, a nine-stage characterization of food waste was done as follows (Garcia-Garcia et al., 2015, 2017):

1. **Edibility**: Based on edibility, waste is classified as edible and inedible:

    i. **Edible Waste:** It includes the food likely to be consumed by humans, e.g., meat, tomato, etc. It also includes the foods comprising edible and inedible parts, e.g., banana which comprises of both the edible part (banana) and the inedible part (banana peel), and an egg, which has an inedible shell.

    ii. **Inedible Waste:** It includes the production byproducts of farms and manufacturing units, e.g., offal's, twigs, bones, peels, etc. Some foods, such as some types of offal, are considered suitable for eating from the biological viewpoint but are not demanded consumption and are therefore considered inedible.

2. **State:** In addition to edibility criteria, waste can be categorized on the basis of the state as eatable and uneatable:

    i. **Eatable:** It includes the waste which is still fit for eating at the time of its disposal.

    ii. **Uneatable:** It includes products whose expiry date has passed and are spoiled and thus unfit for consumption. It also includes the products which have undergone bad processing, e.g., overcooked during processing, having unacceptable burnt flavors, and are thus damaged. Furthermore, it includes a class of products, which are unfit for human consumption but can still be used for animal feed, e.g., products that have fallen from the conveyor belt during the manufacturing process.

3. **Origin:** On the basis of origin, waste is classified as animal-based and plant-based:

   i. **Animal-based:** Waste is said to be animal-based if its origin was from an animal source, e.g., eggs, honey, or was a part of an animal, e.g., meat, bones, offals.

   ii. **Plant-based:** Waste is said to be plant-based if it is of plant origin.
   If the product contains a mix of both animal and plant-based materials, e.g., ready-to-eat meals, then classification must be done on the basis of the predominant ingredient present.

4. **Complexity:** It includes a single product and a mixed product:

   i. **Single Product:** It includes products that are made from only one type of ingredient.

   ii. **Mixed Product:** It includes products that are made from more than one ingredient.

5. **Animal Product Presence:** On the basis of animal product presence, the waste is classified as meat, animal product, and byproduct:

   i. **Meat:** It includes fish.

   ii. **Meat Products:** It includes products produced by the animal such as milk, eggs, honey.

   iii. **Meat Byproducts:** It includes the products that are not intended for human use or consumption, such as offal, hides, skin.

   Meat byproducts undergo various treatments in order to obtain the desired compounds of interest, such as obtaining fat (rendering) to be used in soap manufacture. Therefore, the meat byproducts are debarred for use as animal feed under the EU Regulation No 1069/2009. Also, if the product is from a plant source and is mixed, it must be taken care that the product has not come in contact with the animal-based material in order to avoid contamination and disease outbreaks.

Characterization of Food and Agricultural Wastes

6. **Stage of Supply Chain:** It includes catering and noncatering waste:

   i. **Catering Waste:** The waste generated from food services, i.e., hotels, restaurants, etc., and households are known as the catering waste. As per the UK and EU regulations, 2001 and 2003, respectively, catering waste must not be used as farm animal feed.

   ii. **Noncatering Waste:** It comprises the waste generated in the early levels of the logistic network, i.e., during the cultivation, manufacturing, allocation, etc.

7. **Treatment:** On the basis of treatment, waste is categorized as processed and unprocessed:

   i. **Processed Waste:** It includes the food product, which has the properties alike as the finished product and does not involve any processing requirements.

   ii. **Unprocessed Waste:** It includes the waste which needs to be treated at the time of its management.

8. **Packaging:** It comprises packaged and unpackaged waste:

   i. **Packaged:** It includes the packaged product. If unpacking technology is available for separating the package and food waste, then the product is regarded as unpackaged, or else it is considered as packaged.

   ii. **Unpackaged:** It includes a product that is not contained in packaging material.

9. **Packaging Bio-Degradability:** It includes bio-degradable and non-biodegradable packaging:

   i. **Bio-Degradable Packaging:** It consists of paper, wood, etc. Bio-degradable package refers to the material that can be degraded by microorganisms. This type of packaging waste is suitable for anaerobic degradation in a composting plant.

   ii. **Non-Biodegradable Packaging:** It includes plastic, metal, etc., which is not suitable for AD, composting, or land spreading.

Food waste can also be classified on the basis of the generation of waste in the food logistics network. The waste generated in the food supply network of the fruit and vegetable and animal sector can be divided into five categories (Gustavsson et al., 2011):

1. **Fruit and Vegetable Sector:** Waste is generated in the following segments of the fruit and vegetable sector in the food processing chain:

   i. **Production:** It includes wastage due to mechanical damage at the time of harvesting (fruit picking) or spillage (during threshing).

   ii. **Postharvest Handling and Storage:** It refers to the wastage at the time of handling, storage, and transportation.

   iii. **Processing:** It includes the waste generated during processing operations such as canning, juice production, etc.

   iv. **Distribution:** It is the wastage attributed to the marketing systems such as wholesale markets, retailers, and supermarkets.

   v. **Consumption:** It is the waste produced at the utilization level, e.g., at the household level, at restaurants, cafes, etc.

2. **Animal Sector:** The waste generation in the animal sector takes place at the following segments of the food processing chain:

   i. **Production:** The waste generated during the production in the animal sector is usually due to the death of the animal during breeding.

   ii. **Postharvest Handling and Storage:** In this case, waste is generated as a result of the death of an animal during transportation to a slaughterhouse or due to disapproval at the time of slaughter. The byproducts of animals such as milk are also wasted due to the spillage and spoilage. During transportation, a lot of milk goes to waste in the process of transportation from farm to distribution center and then to the collection center.

   iii. **Processing:** It includes the wastage during the processing of meat, such as during sausage production or during smoking or canning of meat. In the case of milk, the wastage at the time of processing occurs due to faulty milk treatments such as over-heating of milk which results in the burnt flavor of the milk and hence results in the wastage of this unacceptable milk.

iv. **Distribution:** Here, wastage occurs at the time of distribution to wholesale markets, retailers, and supermarkets.

v. **Consumption:** Consumption of waste includes the waste generated at the household level, at restaurants, cafes, etc.

Additionally, food wastage in high-income countries can be categorized on the following basis (Gustavsson et al., 2011):

1. **Quality Standards:** The food is wasted due to the esthetic defects (color, broken, blemished), e.g., carrots are discarded if they don't have a bright orange color. The loss due to quality standards is about 25–30%

2. **Food Manufacture:** An ample quantity of food is wasted in the manufacturing process. Almost <10% of the potato is discarded during the sorting process, 2–12% of fruits and vegetables go to waste in the trimming process. In addition to this, 1–10% of food is wasted in the transportation loss during processing.

3. **Poor Environmental Conditions at the Time of Display:** Poor temperature management at the retail and wholesale stores results in a lot of food being wasted. Some foods are chilling sensitive, i.e., require too cold temperature, and some are heat sensitive, i.e., require too hot temperature. The failure to meet the desired temperature is the main reason for the retail food wastage, which accounts for up to 55%.

4. **Lack of Planning:** Due to the inadequate focus on food wastage, the food ends up being wasted due to negligence and lack of communication or coordination in food storage.

5. **Best-Before-Dates:** Almost 55% of the food in the household of advanced countries like the UK gets wasted because of the expiry of best before and use by date of products.

6. **Leftovers:** The leftover accounts for 42% of the food being wasted (Table 1.1).

TABLE 1.1 Basis of Categorization of Food Waste

SL. No.	Basis of Categorization of Food Waste	References
1.	Type of food	Flores et al. (1999); Malamis et al. (2015)
2.	Nutrient composition: • Carbohydrate content, amount of fat	Russ and Meyer (2004)
3	Chemical composition: • Carbon, Hydrogen, Nitrogen content	Roberts (2015)
4.	Temperature of storage: • Frozen, Ambient	Mena et al. (2011)
5.	Stage of supply chain: • Harvesting stage, manufacturer stage Edibility: • Avoidable, unavoidable, possibly avoidable	Bridgwater and Quested (2013)
6.	Household waste: • Cooked, uncooked • Packaged, unpackaged	Ventour (2008); Quested and Murphy (2014)
7.	Household waste: • Leftover, untouched	Schott and Andersson (2015)
8.	Waste characterization: • Organic crop residue • Catering waste • Animal byproduct • Packaging • Mixed Food waste • Domestic Waste	Lin et al. (2013)
9.	Cause of waste generation: • Transport • Inventory • Motion • Waiting • Overproduction • Over-processing • Defects	Chabada et al. (2012)

Characterization of Food and Agricultural Wastes

TABLE 1.1 *(Continued)*

SL. No.	Basis of Categorization of Food Waste	References
10.	Waste classification: Edibility: • Edible, inedible State: • Eatable, uneatable Origin: • Animal-based, plant-based Complexity: • Single product, mixed product Animal product presence: • Meat, animal product, meat byproduct Stage of supply chain: • Catering, noncatering waste Treatment: • Processed, unprocessed Packaging: • Packaged, un-packaged Packaging biodegradability: • Bio-degradable, non-biodegradable packaging	Garcia-Garcia et al. (2015, 2017)
11.	Food supply chain (F&V, animal): • Production, postharvest handling, processing, distribution, consumption	Gustavsson et al. (2011)
12.	Food wastage in high-income countries: • Quality standards • Food manufacture • Poor state of environmental conditions at the point of display • Lack of planning • Best-before-dates • Leftover	Gustavsson et al. (2011)

1.3 AGRICULTURAL WASTE

According to the United States Environmental Protection Agency (USEPA), agriculture waste refers to the byproduct produced by the rearing, production, and harvest of crops and trees (Sarmah, 2009).

Organization for Economic Cooperation and Development (OECD) defines agriculture waste as waste produced from diverse agricultural operations. It comprises waste from harvest, fertilizers runoff from fields, pesticide residues, field silt and salt, farm waste, waste from slaughterhouses and poultry (United, 1997).

In a generalized way, agricultural waste comprises the residue of agricultural products like fruits and vegetables, meat, poultry, and dairy and crops. The agricultural waste may comprise materials that can offer benefits to the population. However, it is economically less significant compared to the costs of collection, processing, and transportation. The agricultural waste composition varies according to the agricultural activity. The waste can be a liquid, solid, or slurry (Obi et al., 2016).

1.3.1 CHARACTERIZATION OF AGRICULTURAL WASTE

Agricultural waste can be broadly categorized as:

1. **Animal Waste**: It includes manure and animal carcass.

2. **Food Processing Waste**: It refers to waste generated as a result of processing, such as during the processing of maize, only 20% of it is canned, and the other 80% goes waste.

3. **Crop Waste:** It includes sugarcane bagasse, fruit, and vegetable culls, rice stalks, etc.

4. **Hazardous and Toxic Agricultural Waste:** It comprises herbicides, pesticides, insecticides, etc.

1.3.1.1 ANIMAL WASTE

The livestock waste includes the following:

i. **Solid Waste**: It includes the solid content found in the slaughterhouse, such as manure and organic materials.

ii. **Liquid Waste**: It refers to the urine of the animals, water used in the sanitation of slaughterhouses and cages, and water used for bathing animals.

iii. **Gaseous Waste:** It includes the air pollutants viz. H_2S and CH_4 and other odors produced by animals.

Animal waste generation causes a serious threat to the environment—untreated animal waste results in the generation of GHGs. Also, the whiff emanated from the cages as a result of the digestion process of the animal waste results in air pollution. In animal waste, almost 75–95% of the total volume includes water, while the rest accounts for organic and inorganic matter and some species of microorganisms and parasites, which confers to negative environmental effects and spread of diseases to humans (Hai et al., 2010).

Aquaculture waste is another type of animal waste. The waste generated in the majority of aquaculture is metabolic waste. In aquaculture, 30% of the solid that is used as feed becomes waste. The feeding rate in the aquaculture is increased when there is a soar in temperature, which results in the augmentation of waste production. Water flow pattern also affects waste generation. A proper flow of water will lessen the disintegration of fish feces and permit solids to be settled rapidly (Miller and Semmens, 2002; Mathieu and Timmons, 1995).

1.3.1.2 FOOD PROCESSING WASTE

Food industries generate an enormous amount of waste due to food processing. The various waste generating agencies include:

i. Meat Industry: It produces an enormous amount of waste, both solid and liquid such as skin, manure, hooves, blood, urine, etc., along with the offensive odors. Additionally, it results in air pollution by means of the smoke produced due to various processing operations such as rendering, which is an evaporative process and produces a lot of smoke (Jayathilakan et al., 2012).

ii. Milk Industry: Milk processing industry produces hardly any solid waste. However, the wastewater generation from this industry is high. The only solid waste that is produced by this industry is the sludge that is being produced by wastewater purification. A large amount of wastewater is produced in the cleaning and sanitation operations. Wastewater generation results from the cleaning of tanks, storage trucks, and in-line pipe washing and sanitization. A colossal amount of wastewater is also generated in the manufacture of curd, cheese, butter which, however, results in the production of curd particles, whey, and buttermilk. With industrialization taking place, milk production has also increased around 2.8% per annum; the processing of dairy is deemed as an outsized producer of industrial wastewater (Britz et al., 2006). Apart from this dairy industry also contributes to air pollution by emanating GHGs such as CO_2, CO, SO_2 (Kolev Slavov, 2017)

iii. **Fruit and Vegetable Industry:** Fruit and vegetable industry produces two kinds of waste:

    a. **Solid Waste**: such as skin, peel, seed, stones, etc.

    b. **Liquid Waste**: such as juice and wastewater.

The waste acquired from the fruit and vegetable processing industry varies tremendously due to the use of an extensive multiplicity of fruits and vegetables, the wide variety of processes, and the diversity of the product (William, 2005). A massive amount of waste is generated from this industry in the form of skin and peel of the fruit. The skin of the fruits is usually discarded before eating, so the waste generation is more. The inedible portion of the vegetables and fruit accounts for 25–30% (Ajila et al., 2010). The processing of fruits and vegetables produce a sizeable sum of pods, peel, stones, and seeds residues (Babbar and Oberoi, 2014). A lot of waste is derived from the grape and apple industry in the form of pomace, banana, and citrus in the form of peel, mango, and guava in the form of seed and peas in the form of shells (Gupta and Joshi, 2000). The processing waste consists mainly of soluble sugars and fiber. The waste produced is dumped in the landfills, which creates environmental concerns. The liquid waste from this industry, i.e., the wastewater, is generated from fruit washing and equipment cleaning and sanitization. Discarding liquid waste puts in nutrients in excess to the aquatic milieu. This results in uneven nutrient balance, algal overgrowth, and a decrease in oxygen which causes the death of animals viz., crustaceans, amphibians, and fish.

iv. **Oil Industry:** It generates waste in the form of:

    a. **Wastewater:** It comes largely from floor washing, cooling tower, filter press, etc. Wastewater from an oil refinery is acidic and tainted with oil and colloidal particles, which pose

Characterization of Food and Agricultural Wastes 17

a serious threat to the environment; thus, making it imperative to be treated before disposal.

b. Solid Waste: It includes spent earth, spent catalyst, chemical, and biological sludge.

The various processes, such as degumming, result in the generation of lecithin and soap, which is a waste generated from the neutralization process (Pandey et al., 2003).

1.3.1.3 CROP WASTE

It is defined as the waste that is generated from crops; it includes:

i. **Field Waste:** It includes the waste that is present in the field or orchard after the harvest. It consists of leaves, stubble, stalks, pods, etc. This residue is either reinvested in the ground by plowing, or it is burned (Richards et al., 1984). Stubble burning, i.e., the process of setting fire to the crop residue in order to remove it from the field for the sowing of the next crop, has a multitude of deleterious effects. The burning results in the release of gases like CH_4, CO, the volatile organic compound (VOC), and polycyclic carcinogenic aromatic hydrocarbons in the atmosphere, which result in the heavy thick coverlet of smog which has adverse effects on human health and contributes to the air pollution. Besides this, the heat results in the loss of useful microbes and soil fertility by depletion of nutrients (Batra, 2017).

ii. **Process Waste:** It consists of materials left behind once the crop is processed into a valuable resource. It includes husks, seeds, roots, bagasse, molasses, etc. This waste can be put to use as animal feed and also to enhance soil fertility (Richards et al., 1984).

1.3.1.4 HAZARDOUS AND TOXIC AGRICULTURAL WASTE

The growth of the crop also supports the generation of insects and the development of weeds. In order to combat this situation, the farmers encourage the utilization of a range of insecticides and pesticides. After the use of pesticides, most of the pesticide packages are thrown in the fields or ponds themselves, which results in serious environmental concerns such as contamination of the farmland. The issues of food poisoning and food safety also take place due to the presence of pesticide residue in crops. In order to amplify the yield of their produce, some farmers practice the unwarranted use of inexpensive inorganic fertilizers. The surplus use of fertilizer results in some of the fertilizer being absorbed and retained in the soil and a portion of it being passed into the lakes and rivers due to irrigation or runoff. The applied pesticides and insecticides also gain entry into the atmosphere by means of evaporation and hence cause air pollution (Table 1.2; Hai et al., 2010).

TABLE 1.2 Agricultural Waste Categorization

Agricultural Waste Categorization			
Animal Waste	**Food Processing Waste**	**Crop Waste**	**Hazardous and Toxic Agricultural Waste**
• Solid waste	• Meat industry	• Field waste	• Insecticides
• Liquid waste	• Fruit and vegetable processing industry	• Crop waste	• Pesticides
• Gaseous waste			
	• Milk industry		
	• Oil industry		

1.4 GLOBAL SCENARIO OF WASTE GENERATION

An even-handed accord between humans and nature is a prerequisite for peaceful living. With the growing population, urbanization, and economic development, the need for food has also tremendously increased, which has led to an increase in food production and hence an increase in waste generation. Industrialization results in urbanization, and developing countries such as India are in the industrialization phase. The reallocation of people towards urban areas has amplified the per capita solid waste generation. Globally, 7–9 billion a ton of waste is generated per annum (Wilson and Veils, 2015). India is one of the fastest-growing economies in the world, with a 7.30% GDP and an expected growth rate of 10% by 2030. An increase in GDP results in an improved standard of living. All these factors club together and lead to an increased rate of per capita waste generation (Nandan et al., 2017). Globally one-third of food produced for human consumption is lost or wasted, and it accounts for about 1.3 billion tons per year. This also implies that a huge amount of the resources used in food production go in vain. A lot of food is wasted throughout the supply chain. The municipal solid waste generated annually accounts for 2.01

billion tons, and almost 33% of it is not dealt with in a manner, which is safe for the environment. On average, almost 0.11 to 4.54 kilograms/person/day of waste is generated worldwide. High-income countries are said to generate about 683 million tons (34%) of the world's waste. On a per-capita basis, the food wastage in the industrialized world is more than the developing countries. As per an estimate, food waste in Europe and North America is 95–115 kg/year, while as in Sub-Saharan Africa and South/Southeast Asia is only 6–11 kg/year (Gustavsson et al., 2011). Waste generation in high-income countries is projected to increase by 19% daily per capita by 2050. However, in low- and middle-income countries, it is expected to amplify by approximately 40% or more. The waste generation in low-income countries in 2050 is likely to rise by more than three times. Almost 23% of world waste is generated by East Asia and Pacific region and 6% by the Middle East and North Africa region. South Asia, the Middle East, North Africa, and Sub-Saharan Africa are the fastest-growing regions, and waste generation is said to double and triple, respectively, by 2050 (Kaza et al., 2018). The waste generation by 2050 is said to increase by 70% from 2.01 billion tons in 2016 to 3.40 billion tons annually. The per capita food waste in kilograms of selected countries worldwide in 2017 is shown in Figure 1.2 (Ian Tiseo, 2019).

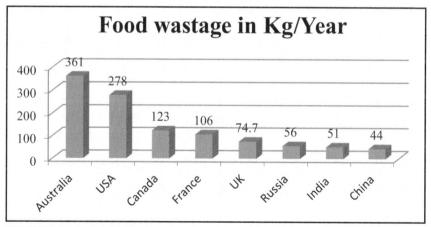

FIGURE 1.2 Per capita, food waste in kilograms of selected countries worldwide in 2017.

Food wastage causes a significant environmental issue. Almost 40% of the waste produced globally is not managed properly and is either dumped or burned (Kaza et al., 2018). According to the World Bank report, almost 10 years back, 0.68 billion tons/year of municipal solid waste was generated, i.e., 2.9 billion urban residents generated about 0.64 kg of MSW/person/day. However, currently, the report estimates that the amount has risen up to 1.3 billion tons per year which implies 3 billion residents produce 1.2 kg of MSW/person/day. By 2025 it is estimated to be about 2.2 billion tons per year, i.e., 4.3 billion urban residents generating about 1.42 kg/capita/day of municipal solid waste (Kaza et al., 2018). India still has an unscientific approach towards waste management (WM). The efficiency of the collection of solid waste is around 70% in India, while in developed countries, the same is almost 100% (Sharholy et al., 2007). A significant sum of the waste is discarded in landfills and city outskirts without prior treatment, which results in adverse environmental impact and is also harmful to the living organisms (Misra and Pandey, 2005). 70% of the world's waste ends up being dumped (Kaza et al., 2018). In high and upper-middle-income countries, 37% of waste is discarded in a landfill and 8% in landfill gas collection systems. However, low-income countries generally dispose of waste in the open. Open dumping of waste accounts for almost 93% in low-income countries and only 2% in high-income countries. The high-income countries have a high percentage of landfill dumping of almost 54%. The Middle Eastern and North African, Sub-Saharan Africa, and South Asian regions explicitly discard more than half of their waste. This poor WM results in water bodies' contamination and drains clogging, which in turn result in causing floods, disease transmission, and harm to animals who unknowingly consume the waste (Eriksson et al., 2005). The waste treatments also pose a threat to the environment. Almost 3–4% greenhouse gas, i.e., 1.6 billion tons of CO_2, is produced from waste treatment which is anticipated to increase to about 2.38 billion tons of CO_2-equivalent/year by 2050 (Monni et al., 2006). Food waste accounts for virtual emissions of about 50%. Additionally, the burning of wastes results in deterioration of air quality which has strong adverse effects on human health (Wiedinmyer et al., 2014).

1.5 CONCLUSION

Food wastage occurs from farm to fork. The categorization of wastes is imperative to minimize and manage waste. The waste produced all through the FSC and at household and farm levels has adverse effects on both

Characterization of Food and Agricultural Wastes

environment and human health. Immense efforts are being taken for the prevention and reduction of food wastage. The increasing carbon footprint and wastage of useful resources will have momentous consequences on sustainability. Besides, the sad plight of hunger pangs cannot be ignored. So, it's an indispensable need of an hour for the scientific community to come at the forefront and advocate their voice on the global podium in order to resolve the matter of food wastage.

KEYWORDS

- **anaerobic digestion**
- **food and agriculture organization**
- **food loss**
- **food supply chain**
- **Sustainable Development Goals**

REFERENCES

Abdel-Shafy, H. I., & Mansour, M. S., (2018). Solid waste issue: Sources, composition, disposal, recycling, and valorization. *Egyptian Journal of Petroleum, 27*(4), 1275–1290.

Additives, F., & Geneva, S. (2015). Food and Agriculture Organization of the United Nations.

Agamuthu, P., (2009). Challenges and opportunities in agro-waste management: An Asian perspective. In: *Inaugural Meeting of First Regional 3R Forum in Asia* (pp. 11, 12).

Agriculture Organization, (2008). *The State of Food and Agriculture 2008: Biofuels: Prospects, Risks, and Opportunities* (Vol. 38). Food & Agriculture Org.

Ajila, C. M., Aalami, M., Leelavathi, K., & Rao, U. P., (2010). Mango peel powder: A potential source of antioxidants and dietary fiber in macaroni preparations. *Innovative Food Science & Emerging Technologies, 11*(1), 219–224.

Assembly, G., (2015). Sustainable development goals. *SDGs, Transforming Our World: The, 2030,* 338–350.

Babbar, N., & Oberoi, H. S., (2014). Enzymes in value-addition of agricultural and agro-industrial residues. *Enzymes in Value-Addition of Wastes, 29.*

Baiano, A., (2014). Recovery of biomolecules from food wastes—A review. *Molecules, 19*(9), 14821–14842.

Batra, M. C., (2017). Stubble burning in North-West India and its impact on health. *Journal of Chemistry, Environmental Sciences, and its Applications, 4*(1), 13–18.

Bridgwater, E., & Quested, T., (2013). Waste & resources action program (WRAP). *Synthesis of Food Waste Compositional Data*, 2012.

Britz, T. J., Van, S. C., & Hung, Y. T., (2006). Treatment of dairy processing wastewaters. *Waste Treatment in the Food Processing Industry*, 1–28.

Buzby, J. C., & Hyman, J., (2012). Total and per capita value of food loss in the United States. *Food Policy, 37*(5), 561–570.

Chabada, L., Dreyer, H. C., Romsdal, A., & Powell, D. J., (2012). Sustainable food supply chains: Towards a framework for waste identification. In: *IFIP International Conference on Advances in Production Management Systems* (pp. 208–215). Springer, Berlin, Heidelberg.

Eriksson, O., Reich, M. C., Frostell, B., Björklund, A., Assefa, G., Sundqvist, J. O., & Thyselius, L., (2005). Municipal solid waste management from a systems perspective. *Journal of Cleaner Production, 13*(3), 241–252.

Flores, R. A., Shanklin, C. W., Loza-Garay, M., & Wie, S. H., (1999). Quantification and characterization of food processing wastes/residues. *Compost Science & Utilization, 7*(1), 63–71.

Food and Agriculture Organization of the United Nations, (2013). *Food Wastage Footprint: Impacts on Natural Resources: Summary Report*. FAO.

Ga, U., (2015). Transforming our world: The 2030 agenda for sustainable development. *Division for Sustainable Development Goals:* New York, NY, USA.

Garcia-Garcia, G., Woolley, E., & Rahimifard, S., (2015). A framework for a more efficient approach to food waste management. *Int. J. Food Eng., 1*(1), 65–72.

Garcia-Garcia, G., Woolley, E., Rahimifard, S., Colwill, J., White, R., & Needham, L., (2017). A methodology for sustainable management of food waste. *Waste and Biomass Valorization, 8*(6), 2209–2227.

Gupta, K., & Joshi, V. K., (2000). Fermentative utilization of waste from the food processing industry. *Postharvest*.

Gustavsson, J., Cederberg, C., Sonesson, U., Van, O. R., & Meybeck, A., (2011). *Global Food Losses and Food Waste*. FAO, Rome, Italy.

Hai, H. T., & Tuyet, N. T. A., (2010). *Benefits of the 3R Approach for Agricultural Waste Management (AWM) in Vietnam: Under the Framework of Joint Project on Asia Resource Circulation Research*.

Ian, T., (2019). *Per Capita Food Waste of Selected Countries Worldwide 2017*. https://www.statista.com/statistics/933059/per-capita-food-waste-of-selected-countriess (accessed on 30 December 2021).

Jayathilakan, K., Sultana, K., Radhakrishna, K., & Bawa, A. S., (2012). Utilization of byproducts and waste materials from meat, poultry, and fish processing industries: A review. *Journal of Food Science and Technology, 49*(3), 278–293.

Kaza, S., Yao, L., Bhada-Tata, P., & Van, W. F., (2018). *What a Waste 2.0: A Global Snapshot of Solid Waste Management to 2050*. The World Bank.

Kolev, S. A., (2017). General characteristics and treatment possibilities of dairy wastewater: A review. *Food Technology and Biotechnology, 55*(1), 14–28.

Kroyer, G. T., (1995). Impact of food processing on the environment—An overview. *LWT-Food Science and Technology, 28*(6), 547–552.

Lin, C. S. K., Pfaltzgraff, L. A., Herrero-Davila, L., Mubofu, E. B., Abderrahim, S., Clark, J. H., & Thankappan, S., (2013). Food waste as a valuable resource for the production

Characterization of Food and Agricultural Wastes

of chemicals, materials, and fuels. Current situation and global perspective. *Energy & Environmental Science, 6*(2), 426–464.

Malamis, D., Moustakas, K., Bourka, A., Valta, K., Papadaskalopoulou, C., Panaretou, V., & Sotiropoulos, A., (2015). Compositional analysis of biowaste from study sites in Greek municipalities. *Waste and Biomass Valorization, 6*(5), 637–646.

Mathieu, F., & Timmons, M. B., (1995). In: Wang, J. K., (ed.), *Techniques for Modern Aquaculture*. American Society of Agricultural Engineers, St. Joseph, MI.

Melikoglu, M., Lin, C. S. K., & Webb, C., (2013). Analyzing global food waste problem: Pinpointing the facts and estimating the energy content. *Central European Journal of Engineering, 3*(2), 157–164.

Mena, C., Adenso-Diaz, B., & Yurt, O., (2011). The causes of food waste in the supplier–retailer interface: Evidences from the UK and Spain. *Resources, Conservation, and Recycling, 55*(6), 648–658.

Miller, D., & Semmens, K., (2002). *Waste Management in Aquaculture* (Vol. 8). West Virginia University Extension Service Publication No. AQ02-1. USA.

Mirabella, N., Castellani, V., & Sala, S., (2014). Current options for the valorization of food manufacturing waste: A review. *Journal of Cleaner Production, 65*, 28–41.

Misra, V., & Pandey, S. D., (2005). Hazardous waste, impact on health and environment for development of better waste management strategies in future in India. *Environment International, 31*(3), 417–431.

Monni, S., Pipatti, R., Lehtilä, A., Savolainen, I., & Syri, S. (2006). Global climate change mitigation scenarios for solid waste management. *VTT Publications: Espoo, Finland, 603*.

Nandan, A., Yadav, B. P., Baksi, S., & Bose, D., (2017). Recent scenario of solid waste management in India. *World Scientific News,* (66), 56–74.

Obi, F. O., Ugwuishiwu, B. O., & Nwakaire, J. N., (2016). Agricultural waste concept, generation, utilization, and management. *Nigerian Journal of Technology, 35*(4), 957–964.

Östergren, K. (2014). FUSIONS definitional frameword for food waste (FP7-rapport).

Overcash, M. R., Humenik, F. J., & Miner, J. R. (1983). *Livestock Waste Management. Volume I*. CRC Press, Inc.

Pandey, R. A., Sanyal, P. B., Chattopadhyay, N., & Kaul, S. N., (2003). Treatment and reuse of wastes of a vegetable oil refinery. *Resources, Conservation, and Recycling, 37*(2), 101–117.

Peano, C., Migliorini, P., & Sottile, F. (2014). A methodology for the sustainability assessment of agri-food systems: an application to the Slow Food Presidia project. *Ecology and Society, 19*(4).

Quested, T., & Murphy, L., (2014). Waste & resources action program (WRAP). *Household Food and Drink Waste: A Product Focus*.

Richards, B. K., Wafter, M. F., & Muck, R. E., (1984). Variation in line transect measurements of crop residue cover. *Journal of Soil and Water Conservation, 39*(1), 60, 61.

Roberts, D., (2015). Characterization of chemical composition and energy content of green waste and municipal solid waste from Greater Brisbane, Australia. *Waste Management, 41*, 12–19.

Russ, W., & Meyer-Pittroff, R., (2004). Utilizing waste products from the food production and processing industries. *Critical Reviews in Food Science and Nutrition, 44*(1), 57–62.

Sands, P., (1992). The United Nations framework convention on climate change. *Rev. Eur. Comp. & Int'l Envtl. L., 1*, 270.

Sarmah, A. K., (2009). Potential risk and environmental benefits of waste derived from animal agriculture. In: Geoffrey, S. A., & Pablo, A., (eds.), *Agriculture Issues and Policies Series-Agricultural Wastes.*

Schott, A. B. S., & Andersson, T., (2015). Food waste minimization from a life-cycle perspective. *Journal of Environmental Management, 147*, 219–226.

Sharholy, M., Ahmad, K., Vaishya, R. C., & Gupta, R. D., (2007). Municipal solid waste characteristics and management in Allahabad, India. *Waste Management, 27*(4), 490–496.

Stuart, T., (2009). *Waste: Uncovering the Global Food Scandal.* WW Norton & Company.

United, N., (1997). *Glossary of Environment Statistics, Studies in Methods.* United Nations New York, NY.

Ventour, L., (2008). *Food Waste Report—The Food We Waste.* Waste & resources action program (WRAP): Banbury, UK.

Wiedinmyer, C., Yokelson, R. J., & Gullett, B. K., (2014). Global emissions of trace gases, particulate matter, and hazardous air pollutants from open burning of domestic waste. *Environmental Science & Technology, 48*(16), 9523–9530.

William, P. T., (2005). In: John, W., (ed.), *Water Treatment and Disposal* (p. 9). Great Britain.

Wilson, D. C., & Velis, C. A. (2015). Waste management–still a global challenge in the 21st century: An evidence-based call for action. *Waste Management & Research, 33*(12), 1049–1051.

World Biogas Association, (2018). *Global Food Waste Management: An Implementation Guide for Cities.* Full Report.

CHAPTER 2

Food Processing Byproducts: Their Applications as Sources of Valuable Bioenergy and Recoverable Products

UFAQ FAYAZ,[1] IQRA BASHIR,[1] JIBREEZ FAYAZ,[2] OMAR BASHIR,[1] SOBIYA MANZOOR,[2] TAWHEED AMIN,[1] and SHAKEEL AHMAD BHAT[3]

[1] *Department of Food Technology and Nutrition, Lovely Professional University, Punjab, India*

[2] *Department of Food Technology, Islamic University of Science and Technology, Awantipora – 192122, Jammu and Kashmir, India*

[3] *College of Agricultural Engineering and Technology, Sher-e-Kashmir University of Agricultural Sciences and Technology of Kashmir, Shalimar, Jammu and Kashmir, India*

ABSTRACT

Agri-refuse can be generated through different sources, mostly coming from livestock, plantation, and aquaculture. In addition to wasting tons of relevant biomass resources, inaccurate disposal of agricultural wastes also results in environmental pollution. The significant step in protecting the environment, energy structure, growth of agriculture, and economy is attributed to the recycling and utilization of agricultural wastes. The agrarian straw and livestock excreta is being considered as potential resource. The 3R-strategy of waste management (WM)

Integrated Waste Management Approaches for Food and Agricultural Byproducts. Tawheed Amin, PhD, Omar Bashir, Shakeel Ahmad Bhat & Muneeb Ahmad Malik (Eds.)
© 2023 Apple Academic Press, Inc. Co-published with CRC Press (Taylor & Francis)

is used to convert the wastes for a wide range of applications. Since agri-refuse are high in bioactives, which can thus be employed in the manufacturing of enzymes, antibiotics, animal feed, biofuels, and other useful substances. The well-planned exploitation of animal byproducts can increase the existing cost and dearth of feed materials, which have soaring competition between humans and animals. Meat and poultry processing refuse can be recycled as raw materials or can be turned into products of increased value. Byproducts like liver, lung, kidney, brains, spleen, and tripe have good nutritive value. In addition, the dairy byproducts, which are high in nutritive value, also find utility in both food and non-food industries. Moreover, in the era of modernization and urbanization, recycling, and utilization of agri-refuse has become a prerequisite.

2.1 INTRODUCTION

According to the European Union (EU) Waste Directive, whichever matter or entity that possesses or disposes of or aims to throw away or is necessitated to remove, is termed as "waste" (Millati et al., 2019). Food waste primarily consists of unprocessed or cooked food material which is discarded on the food supply chain (FSC) and cannot be used (Burange et al., 2016). Food and Agriculture Organization (FAO) defines food squander as the victuals lost taking place on the road to the conclusion of the nourishment chain connected through the conducts of end-users and dealers, and it is spaciously acknowledged (Parfitt et al., 2010). Despite, wastes from agriculture are characterized like the remains from the budding and handling of crude agrarian goods such as natural products-meat, fresh whole fruit and vegetables, poultry, dairy products, whole grains, and crops. These are the non-product capitulates of manufacturing and preparing of agrarian items that possibly will enclose substances that have the potential to benefit humans, however, whose money-making standards are fewer than the expenditure of assembling, shipping, and preparing for valuable usage (Obi et al., 2016).

The agrarian nourishment items have an amplifying requirement for briskly expanding populace of the world; the rise in populace has been observed as 2.50 billion in 1950s and 6.90 billion in 2010, which may perhaps rise to 9.15 billion in 2050s (Alexandratos and Bruinsma, 2012).

Food Processing Byproducts

In the pre-utilization process of wastes, about one-third of the agri-food items are transferred to wastes such as collecting, storing, wrapping, shipping, and post utilization, i.e., cooked food and leftover food on the consumer's platter (FAO, 2016). Additionally, reported by the Swedish Institute for Food and Biotechnology (Gustavsson et al., 2011), the magnitude of agri-food wastes has been approximated to be 1.30 billion for each year. Furthermore, in 2009 FAO of the United Nations figured that about 32% of the entire food/edibles delivered over the world was vanished or squandered (Buzby and Hyman, 2012). It is also accounted that about 870 million individuals are severely underfed/starved, roughly one-third of the nourishment created for humans, i.e., 1.30 billion tons annually, is ravaged worldwide (Kojima and Ishikawa, 2013). Agricultural production and agro-industrial processing practices produce an elevated quantity of byproducts and squanders.

Above half of fresh fruits account for byproducts like cuttings, peels, shells, bagasse bran, stems, and seeds which can have nourishment or useful substance greater than the end item (Ayala et al., 2011). Wastage of food commodities is also due to the harm caused during shipping, storage, and handling. It is reported that in the recent past, because of the developing prevalence of fruit juices, nectars, solidified, and minimally prepared items, there has been an increase in the production of byproducts and wastes (Leon et al., 2018). The horticulture and agro-industrial divisions deliver huge amount of remains or derivatives that comprise of diverse varieties of natural mixes which includes creature squander (dung, animal carcasses), food handling squander (in which 80% of maize disposed as refuse and 20% is tin canned), harvest squander (dry pulp, corn stalks, drops, and culls from fruits and wastes, sugarcane, and pruning) perilous and lethal agrarian squander (germicide, bug sprays, defoliant, etc.) (Figure 2.1). Food wastes also include wastes from food hubs, restaurants, bars, and furthermore from food generation services which largely entails second-hand cookery oil, food bits and pieces, leftovers, and perished or substandard food. Above 1 billion tons of nourishment squander is yearly produced around the world, the majority of which is conceivably achievable for recycling (Luque and Clark, 2013).

Large-scale pollution of land, water, and air is owing to unorganized degradation of wastes from agricultural, agro-industrial, and food sources. Diverse greenhouse gases (GHGs) discharged from solid throw-away dumps too adds to global warming. Bulky solids, airborne contaminants, and wastewater takes into account waste products from food processing amenities. All these grounds for harsh pollution troubles and are focus to increasing environmental guidelines in nearly every country. By and large, the wastewater is the main alarm because food processing procedures grip an amount of unit operations, for instance, washing, evaporation, removal, and filtration. The method of wastewater operations more often enclose lofty concentrations soluble organics like carbohydrates, lipids, and proteins and suspended solids, that are classified as hard disposal hitch (Hansen and Cheong, 2019). The dumping of food squander is carried out through normal techniques for instance composting, disposing, burning, and anaerobic absorption. Owing to the large amounts of moisture content within the FW strategies, for instance, ignition create much detrimental gasses and composting directs to liberate leachates which lines the procedure ecologically hazardous (Cekmecelioglu et al., 2005). Besides, traditional methods of disposing refuse (such as dumping, burning, and composting) which are not appropriate for food squanders as it is power-intensive, produces hazardous gases (like methane), leachates, and bad smell with major concern to survival. The carbon to nitrogen ratio of squander for the process of composting is cautiously optimized to 25:10 and 35:10 in order to attain the compost of excellent microbial activity and ultimately a compost of superior quality, which might or else show the way to quite a few environmental problems (Burange et al., 2016).

Waste from foods and their derivatives after proper utilization can be used as crude materials or can be added to foods as additives, which creates financial growth of firms, confers to lessen the food crisis, emerges as the aftermath for positive well-being and would shrink the ecological implications that produce disorganization of squander. To achieve zero wastes, industries at present adopt modern technologies, where the waste generated can be employed as unprocessed material for the manufacturing of novel commodities (Leon et al., 2018).

To safeguard the surrounding; scientists are constantly seeking substitutes to handle solid waste. Decreasing of wastes obtained from food and agriculture, and to make use of them in valuable products, have the scope of amplifying the FSC. The discussion wrap together the magnitude and features of waste distinctiveness and WM and possible exploration of waste.

Food Processing Byproducts 29

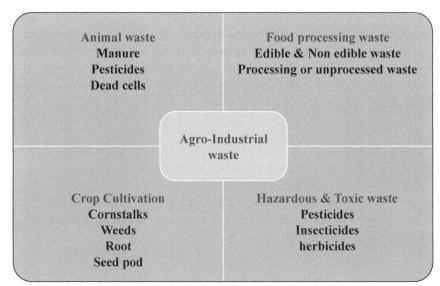

FIGURE 2.1 Different types of wastes.

2.2 HIERARCHY OF AGRICULTURAL AND FOOD WASTE MANAGEMENT

The waste outline order deems food and agricultural waste as primary ecological concern and devise work plan to accurately exploit and administer the food waste, for instance, gathering of biowaste discretely from the other squanders, and the exploitation of biowaste as manure or digestate, beside with the foreword of innovative move toward up toes for the making of priceless and environmental substance from the agricultural and food waste (AFW) (Ankush et al., 2020). The pecking order of waste organization is decreasing the utilization of resources and recycling them to be the most environmentally. Basic drop begins with lowering the magnitude of waste produced and using again materials in order to put a stop on them from piercing the waste stream (Bhada and Themelis, 2008). The idea of decreasing waste lessens the magnitude and adverse effects of waste production by decreasing the amount of squanders, refuse can be reused with easy refinings and salvaging the refuse by treating it as a reserve to manufacture the identical or customized products. This is typically called '*3R*.' A few waste products can be devoured by reusing the same asset for manufacturing of diverse or the same item. Waste reutilization equalizes harvesting of new or same products, which leads to utilization of fresh source and refuse shrinkage. On the whole, the 3Rs alone or in conjunction helps in the utilization of new resources, augments to the previously utilized resources and very prominently reduces the waste extent and hence its adverse impacts (Obi et al., 2016).

The 3R's agro-waste opportunities include:

- Heat production: This practice has 92% as thermal yield (EIA, 2008). Agricultural remainders are burnt as fuel.
- Production of cellulosic ethanol as a biofuel.
- Biogas production as a proxy for cooking gas as an alternative to fuelwood (in rural villages) and so that requirement for cooking biogas in urban areas can be met.

2.3 CONVENTIONAL FOOD WASTE MANAGEMENT TECHNOLOGIES FOR OBTAINING AND UTILIZING BYPRODUCTS

2.3.1 ANIMAL FEEDING

Crop remains have elevated fiber content and have small content of protein, starch, and fat. As a result, the conventional way of amplifying the livestock production by enhancing fodder and field with grains and protein concentrate may not meet the potential of meat protein requirements. Exploiting grain and protein for human food will battle with such use for forage. By using residues to feed domesticated animals these situations may be evaded (Hussein and Sawan, 2010). Under well-managed conditions, substitutes such as landfilling or anaerobic digestion (AD) are not distinctly beneficial as that of animal feeding. Advantages mainly appear from the fractional imitation of normal feed, the manufacturing/ making of which has momentous impact on environment and health. In countries like Japan and South Korea, local laws promote using AFW to feed animals and therefore, the segments of food waste recycled as forage are inflated to 35.90% and 42.50%, respectively. These applications are firmly legalized in these countries; although currently in the EU it is prohibited to utilize food refuse as forage (Salemdeeb et al., 2017).

Food Processing Byproducts

2.3.2 LAND FILLING

Its remarkably the most widespread and ordinary method to discard solid debris in spite of its environmental impact. Landfilling is described as the throwing away, crushing, and infusing the embankments with squander at suitable places (Arvanitoyannis, 2008) and comprises four ordinary phases:

- Hydrolysis/aerobic degradation;
- Hydrolysis and fermentation;
- Acetogenesis; and
- Methanogenesis.

Carbon and supplements that invades the landfills as agro squander result in the discharge of vapors (NH_3, CO_2, CH_4) and fluids (leachate) or get accumulated in the landfill. Most of the power gets engrossed in the collection of refuse and its transference to landfills, transference of leachate to remote areas in order to treat leachate according to regulatory dispensing regulations. After discharge, impurities get concentrated, which may surpass the local ambient water quality standards for surface or groundwater, which should be returned within the normal ecological standards by diluting it with large amounts of water (gray water footprint) (Kibler et al., 2018).

Methane from landfills produced by the anaerobic biological degradation of the dumped organic matter is the third major anthropogenic provenance of atmospheric CH_4 emissions, accounting for around 800 million tons of CO_2-equivalent (Breeze, 2018). The gathering rate of landfill gas is commonly less than 60% in developed countries. USEPA evaluated that the entire anthropogenic emission of CH_4 was 282.6 million tons in 2000, in which 13% (36.7 million tons) was because of landfill emissions (Ren et al., 2017).

2.3.3 ANAEROBIC DIGESTION

In agricultural and industrial waste treatment, AD has been pragmatically employed and acknowledged as a cost-effective method for food and agricultural waste disposal. This process has been broadly applied for the treatment of waste in the European Union and developed Asian countries from 2006 onwards (Abbasi, 2012). This has been established as a popular technology meant for the treatment of slush and manure produced in the refuse water process plants (Mata-Alvarez et al., 2000; Moral et al., 2018). Furthermore, AD is used for reducing food and agricultural waste, which involves mortification of waste without oxygen. The CH_4 present in the biogas is the most beneficial product produced during the AD, but the digestate attained as byproduct of absorption procedure may additionally be ratified, either simply as cattle floor covering or as substratum in soil reforms, or by ensuing composting or vermin composting processes to attain fertilizers (Esparza et al., 2020). The produce acquired during AD is biogas like methane (CH_4), ammonia (NH_3), and carbon dioxide (CO_2) in conjunction with an aqueous product like fertilizers. This process entails the subsequent steps (Trivedi et al., 2020):

- Hydrolysis step;
- Acidogenesis step;
- Acetogenesis step; and
- Methanogenesis step.

Organic substances are putrified, which are categorized into four levels. Initially, polymer organic material in solids is fractured into soluble simpler units along with the conversion from carbohydrates, proteins, amino acid, and long-chain fatty acid and fat to sugar, respectively, and this process is known as hydrolysis. Most of the researchers have suggested that hydrolysis level is regarded as the rate restricting step for multiplex unprocessed matter mortification because of the manufacturing of unpleasant volatile fatty acids or poisonous byproducts (complex heterocyclic compounds) (VFAs) (Yuan and Zhu, 2016). In the next move, known as acidogenesis, monomers mortify into short-chain fatty acid together with VFA; actic acid, pyruvic acid, acetic acid, formic acid. And next, in the process of acetogenesis, digestion of acids like lactic acid and pyruvic acid occurs into acetic acid and hydrogen. In the final stage, hydrogen, and acetic acids are converted into CH_4 by methanogenic archaea, and this process is called methanogenesis, (Ren et al., 2017).

The composition of the typical gas produced is: methane: 50–70%; CO_2: 25–45%; N_2: 0.5–3%; H_2: 1–10% with traces of H_2S; and the value of heating the gas is in the range of 18–25 MJ/m^3. The handling and disposal of large poultry, swine, and dairy waste is made possible by AD and consequently minimizing the odor snag. It neutralizes the waste and the digestion sludge is comparatively odor-free and still retains the fertilizer value of the original waste (Obi et al., 2016).

2.3.4 PYROLYSIS

Pyrolysis has been regarded as an appealing substitute for dumping of agrarian squander because the technique can metamorphose this special biomass resource into granular charcoal, non-condensable gases and pyrolysis oils, which could equip beneficial energy and chemical products owing to their elevated calorific value (Trninic et al., 2016). In pyrolysis systems, agricultural waste is heated up to a temperature of 400–600°C in the absence of oxygen to vaporize a segment of the material, leaving behind char. This is believed to be an advanced technology practice for using the agricultural wastes. They are used for the manufacturing of chemicals from agricultural waste as well as for energy recovery (Obi et al., 2016). The biomass obtained from pyrolysis oil (bio-oil) could be used as feedstock for producing hydrocarbons that could be easily combined into existing petroleum refineries or potential bio-refineries (Kim et al., 2013). The produced gas is able to be used as a fuel for synthetic natural gas production or used for production of liquid fuels (e.g., methanol, ethanol, gasoline, aviation fuel and diesel fuel). The charcoal can be used as an active carbon, as a reducing agent in the metallurgical industry (e.g., to smelt metal ores), as fuel in households (e.g., as a barbecue charcoal), as an adsorbent or as a raw material for the production of chemical compounds (carbon disulfide (CS_2), calcium carbide (CaC_2), silicon carbide (SiC), sodium cyanide (NaCn), as a soil fertilizer, carbon black, various pharmaceutical compounds, etc.) (Trninic et al., 2016; Becidan et al., 2007; Yaman, 2004). Pyrolysis of biomass is still at the stage of being developed for marketable use, and needs to master a large amount of technological and financial barriers to take part with conventional fossil fuel-based procedures (Jahirul et al., 2012), despite various projects have been approved (Trninic et al., 2016).

2.3.5 COMPOSTING AND VERMICOMPOSTING

Aerobic mortification of organic substances occurs due to the activity of microbes such as bacteria, fungi, and protozoa into comparatively reliable products (Banks and Wang, 2006). Under aerobic conditions, the in processed matter is transformed into CO_2, NH_3-N, or compound intractable substances which are frequently called humic. Compost can be utilized as soil redress, which seizes C (Hansen et al., 2016), annuls manufacturing, and utilization of chemical fertilizers, and increases soil moisture withholding (Doan et al., 2015), which decreases the drenching necessities. More newly, vermicomposting of food refuse for the fabrication of bioinoculants has been build up (Rorat and Vandenbulcke, 2019), which embodies the stabilization of organic refuse because of the synergism between earthworms and microbes. The final product (vermicompost) obtained has high nutrient availability in soil, which is thought to possess better properties than its traditional counterpart (Suthar, 2009). Vermicomposting is a bio-collectively mineralizeun processed refuse as substratum and transform them into nutrient-rich unprocessed dung (Pramanik et al., 2011). Vermicompost can be employed for soil conditioning, while as earthworms finds its application in medicine, as nourishment in fish tarns and other technologies like vermin-filtration and vermi-remediation. Vermicompost (which is comprised of nutrients, humic acid and other nourishing hormones) is extensively used as unprocessed fertilizer in chemical-free cultivation. Thus vermin-composting is advantageous over chemical fertilizers and is beneficial for crops. (Dominguez and Edwards, 2011; Manyuchi and Phiri, 2013).

Its a technology that permits the supervision of the refuse unprocessed material, and as a result it is appropriately used for food and agri-waste. Ecological benefits occur from less release of GHGs and leachate generation collated to landfilling.

Vermicomposting practice has been used as one of the nonpolluting technologies for refuse organization as various methods have been performed on stabilizing refuse, produced due to various agrarian activities and as well as from other industries. Earthworms, vermicompost, and vermiwash are yields of vermicomposting. Vermicomposting has also been also used as a cost-efficient practice for refuse water dealing. Vermi filtration helps in the management of sewage water treatments. Huge surface area and enhanced action of cations makes vermicompost an appropriate low-cost organic adsorbent (Dey et al., 2017; Kumar et al., 2015). Consequently, vermicompost is also utilized for eliminating heavy metals and dyes from refuse water (Pereira et al., 2014). Research is showing that in addition to vermicompost, vermicomposting treatment produces another environment efficient byproduct in liquid form. Prospective of these vermicomposting obtained liquids has been shown in numerous studies raised above. The chemical-free method of preparation for these liquids

Food Processing Byproducts

is in support as a way of producing eco-friendly biofertilizer. One of the benefits of vermicomposting derived liquids is that nutrients present are entirely soluble in water because of its processing method. Hence, it has the potential to be used as a foliar fertilizer. To be used as foliar spray, the nutrient should be water-soluble with low salt concentration till it causes scorching on leaves when sprayed. From the outline of studies carried out with vermiwash, vermicomposting leachate and vermicompost aqueous extract, these biofertilizers depicted guaranteed effect in various dilutions. For this reason, vermicomposting derived from liquids can be potentially used as foliar fertilizer (Quaik and Ibrahim, 2013).

2.3.6 THERMAL TREATMENTS

Due to the high moisture content of food waste, energy required for combustion/ ignition is extensive. Hence AFW is generally combusted where energy for evaporation is provided by other incendiary refuse. Electricity is generated by the heat produced by incineration (Kibler et al., 2020). Absolute combustion of agrowaste "consists of the swift chemical reaction (oxidation) of biomass and oxygen, the liberation of energy, and the immediate development of the final oxidation yield of organic matter-CO_2 and water" (Klass, 2004). The energy is typically dissipated in the form of radiant and thermal energy if only the oxidation occurs at a satisfactory rate; the quantity of which is a corollary of the enthalpy of burning of the biomass (Obi et al., 2016). Combustion produces a solid deposit, ash, formed by a fraction of the inorganic portion of the solid fuels, which is usually dumped in land, even though the exercise has been reinforced to use embers as raw/ building material (Dhir et al., 2018). Of all the processes, combustion is the superior technology that has been utilized to metamorphose agri squander into energy or fuels, accounting for greater than 95% of all biomass energy consumed (Klass, 2004).

2.3.7 WASTE VALORIZATION

Above 1 billion tons of food waste is yearly produced universally, the majority of which possibly will be viable for recycling (Luque and Clark, 2013). Nevertheless, the ordinary above-mentioned dumping techniques mostly explain to appoint as conventional customs for food squander execution. Intriguingly squanders processed from the agri-food commerce comprise a forfeiture of beneficial biomass and other supplements, as these squanders have the prospective to embellish into beneficial products or even can be used as raw products for various industries. Suitable utilization of agri-food squander and byproducts as crude supplies or nourishment added substances, might potentially give rise to economic increment to the company, will lead to reduction in health problems, cause increase in favorable health effects and would diminish the ecological burden that causes the mishandling of the squander. Currently, the industries are focused in concoctions to produce zero wastes, where the waste produced by one may be utilized as the raw material by the other for the manufacturing of new products (Leon et al., 2018).

2.3.8 BIODIESEL MANUFACTURING FROM AGRI-FOOD WASTE

Biodiesel which may be defined as a monoalkyl ester of fatty acids or fatty acid methyl ester, is an inexhaustible and clean-blazing fuel substitution for crude oil, making biodiesel production a developing business (Karmee and Lin, 2014). Utilization of biodiesel from the inexhaustible sources like food refuse and cooking oil refuse which is a nonhazardous, eco-friendly, low emissive, recyclable, carbon unbiased (Yaakob et al., 2013). At present, it is one of the excellent substitutes for fuels. In EU and the UK, biofuel is manufactured through rendered fats (Lin et al., 2013). Biofuel can also be prepared from food wastes employing various steps combining fermentation of microalgae and transesterification of enzymes (Li and Yang, 2016). The vital machinery engaged in the production of biodiesel contains:

1. **Transesterification:** Triglyceride, the main constituent of vegetable oil reacts with ethanol in the presence of alkaline catalyst in a process called "trans-esterification." This process converts triglycerides to diglyceride then monoglyceride, and ultimately to glycerol.

2. **Side Reaction One (Saponification Reaction):** If vegetable oil, holding free fatty acids (FFA) will counter with similar alkalines catalyst there is formation of soap and water. The main disadvantage of this reaction is the utilization of catalyst and surged

hardship in segregation course, which causes increase in cost associated with the production. Furthermore, production of water during the reaction hampers the activity of catalysts.

3. **Side Reaction Two (Hydrolysis Reaction):** Moreover, the building of water either from V.O or emerged in the soap formation reactions, causes hydrolyzation of triglycerides to build more FFAs.

4. **Esterification:** Vegetable oil is preprocessed with acid catalysts which lead to the genesis of FFA esters (Biofuel), which eliminates the shaping of soaps because of the reaction of FFAs with the homogenous alkaline catalyst. This reaction is beneficial when the raw material possesses elevated levels of FFAs (esterification of FFA to make FFA esters). However, the aforementioned reaction is dilatory than the alkaline catalyzed transesterification reaction (Gnanaprakasam et al., 2013).

The novel technologies developed in the recent past has facilitated the manufacturing of biofuel from reuse fried oils, evolving into a product whose quality is analogous to that attained from virgin vegetable oil biodiesel (Girotto et al., 2015). According to Canakci (2007), yearly making of greases, oils, and animal fats as of eating places in America might possibly swap beyond 5 Ml of diesel fuel if assembled and transformed to biofuel. Pre-used cooking oil (C.O) needs a round of pre-processing steps in order to get rid of solid contaminants and to decrease the concentration of FFAs. The pre-treatment practices include centrifugation, washing, acid esterification and flash evaporation. The culminating ester outcome might be equal to 80% (Yaakob et al., 2013). These outcomes are anticipated to embolden the government and private divisions to upgrade the assemblage and recycling of already utilized C.O for the manufacturing of biodiesel (Girotto et al., 2015).

Waste lipids can be mixed and watered down with petro-diesel for straight utilization in engines; trans-esterifying the lipids into biodiesel is an essential spot with lots of activities happening. Transesterification can be catalyzed by alkaline, acid, or enzyme. Moreover, the said catalysts are obtained from food squanders. Food squanders can be treated and utilized as a catalyst which encompasses bone, ash, and shell, which can help in the release of calcium oxide (Li and Yang, 2016). Tan et al. (2015)

conducted a study in which he utilized calcinated egg casings as mixed catalyst for the making of biofuel by employing squander cooking oil as raw material. They were described to have an improved catalytic action because egg casings are much permeable and possess a huge area than the shells of mollusk. In order to decrease the quantity of working reactants as saponification and to decrease the catalyst impediment, pretreatment of oil-linked acid esterification is required, but the pretreatment step must be optimized owing to the varied composition of different feedstocks (De Almeida et al., 2015). To improve the product yield transesterification method can be benefited utilizing new technologies which include ultrasound and microwave. To avoid saponification, the structure should be free of water during transesterification procedure (Lin et al., 2013).

Besides canola oil and soybean oil, cooked oil squanders are the foremost bases for the manufacturing of biodiesel. During transesterification method, the metamorphosis of biofuel from cooking oil squander is accomplished with the aid of lipolytic enzymes like Lipozyme TL IM and Novozym435 (Lee et al., 2013). Biofuel can be produced from the minimum work of feedstock like palm unsaturated fat distillate, which is an outcome of the palm oil industry (Lokman et al., 2014). Along with lipases, enzymes from microbes are utilized for the manufacturing of biofuel. Silica-gel matrix is crippled with mixed lipases derived from *Rhizopus oryzae* and *Candida rugosa*, for the manufacturing of biofuel from soybean oil. Employing this methodology, inflated levels of biodiesel was attained and the surrounding substance can be again used up to 30 cycles (Kumar et al., 2020). Food refuse can be transformed into fatty acids and biodiesel due to the transesterification of microbial oils generated by several oil-producing microorganisms (Yaakob et al., 2013; Pleissner et al., 2013). Here the food refuse acts as a substrate for oil-producing microorganism's culture to amass oil with or ensuing to transesterification. Oils from microorganisms can be utilized as an alternative for oils obtained from plants due to their comparable composition of fatty acids, and these oils can be manufactured by several strains of yeast or microalgae. The study of Dr. Carol S.K. Lin's group in the School of Energy and Environment at the City University of Hong Kong (Pleissner et al., 2013) has revealed that the power of FW hydrolysate (FWH) as the only provenance of supplement in microalgae development for biofuel manufacturing. Hydrolyzing the fungus by (FAN), *Aspergillus oryzae* and *Aspergillus awamori*, phosphate, and glucose were regained from food wastes which can be utilized as culture media for the two heterotrophic microalgae, namely *Chlorella pyrenoidosa* and *Schizochytrium mangrovei*. Fine growth of *C. pyrenoidosa* and *S. mangrovei* was obtained by using FWH directing to the manufacturing of 10–20 gram biomass. The two biomass strains were loaded with fats, SFA and (PUFAs), and were stated to be appropriate for the manufacturing of biofuel. The important poly-unsaturated fatty acids (PUFAs), α-linolenic acid (ALA), Docosahexaenoic acid (DHA) and protein content in biomass possess the prospective of being utilized as nourishment and its resources.

Agro-industrial wastes and agrarian processed waste such as coffee ground, waste vegetable oil (received from households, food hubs and agri-food industries, etc.), rice bran and deoiled cakes of both edible and non-edible oil seeds are possible feedstocks for biofuel manufacturing, owing to the remaining oil available in the cakes. Green solvents as dimethyl ether (DME) are used to extort bio-oil from spent coffee powders. The finest choice to remove oil from coffee grounds is DME that contains oil content of about 16.80%, which validates its strength for biofuel generation by the transesterification method (Sakuragi et al., 2016). Global evaluation specifies generation of about 38.50MMT of rice bran from 482 million metric tons of rice. About 17.50% is the approximate percentage of oil content in rice bran (Pattanaik et al., 2019) and 8% FFA content is present in crude rice bran oil and has huge prospective for biodiesel production. Rice bran oil is prepared by two-step pretreatment processes due to existence of high free fatty acid content. At first, in the existence of sulfuric acid and methanol, acid-catalyzed esterification process takes place, which leads to the formation of rice bran oil of low quality accompanied by trans-esterification. Biofuel produced in this process was monitored and ranged from 60–85% with differences in the process parameters which include methanol percent, catalyst percent, and reaction time (Kattimani et al., 2014). Biofuel can also be manufactured using solid acid catalyst, which is advantageous over mineral acids or base catalysts, because of its variant like characters (Jitputti et al., 2006). An output of about 98.10% biodiesel have been monitored for 4 weight percentage of catalyst (K_2CO_3 altered zeolite) (Bani et al., 2018). Likewise, inflated value of biofuel yield (92%) is attained from 40% FFA RBO by employing chlorosulfonic acid changed zirconia during immediate transesterification and esterification processes (one-step response) (Zhang et al., 2013).

It has been discovered that biofuel against edible oilseed cake was similar to deoiled olive oil cake. On an average, the oil content present in olive oil cake is 13.75% and the deoiled cake has been employed as the initiating matter for the manufacturing of biofuel in the course of a 2 way process. Initially, olive oil cake includes FFA of 24.50%, during esterification reaction, it is reduced to 0.52%. By means of transesterification reaction the biofuel output fluctuates from 40% to 65% (Al-Hamamre, 2011). Non-edible oil seeds as karanja, jatropha, simarouba, cottonseed, neem, etc., have been mainly established as the convenient materials for manufacturing of biofuel. Nonetheless, the practice adopted intended for extracting out of oil from the said seeds has a major repercussions on the output of biofuel. Despite the fact that many researchers have built up diverse capable oil removal methods from non-edible seeds, not a single oil extraction procedure has attained a cent percent output (Sakuragi et al., 2016). Less oil content in cake is additionally suitable for the manufacturing of biologically CNG and decoction of value-added products like amino acids and proteins (Chandra et al., 2012; Sanchez-Arreola et al., 2015). Well-planned use of deoiled cake has an affirmative effect on the economy of biofuels.

2.3.9 PHENOLIC COMPOUNDS FROM AGRI-FOOD BYPRODUCTS

Phenolic complexes comprise a huge and varied set of minor metabolites obtained from phenylalanine and tyrosine and are generally doled out in kingdom Plantae. Depending on their fundamental carbon structure, phenolic complexes can be distributed into diverse groups like C_6H_6O, $C_9H_8O_3$, $C_7H_6O_3$, $C_8H_8O_2$ and additional phenylpropanoids as well as HT and CT (proanthocyanidins) (Shahidi et al., 2019). These arrangement of these substances might vary from elementary phenolic molecule to complex large atomic polymer mass and have an aromatic loop carrying one or more hydroxyl clusters. Antioxidant activity of phenolic complexes is dependent on its configuration, especially on the nature of replacements on aromatic rings and in quantity and sites of the hydroxyl groups. Chief bases of phenolic compounds in man's diet include fruits, vegetables, and beverages (Balsundaram et al., 2006).

Phenolic compounds demonstrate a broad variety of physical characteristics, for instance, anti-thrombotic, anti-microbial, anti-allergenic, anti-inflammatory, anti-atherogenic, antioxidant, heart defensive and vasodilatory impacts (Manach et al., 2005). Hence, utilization of foods from plants is influential to support well-being and also decrease in ailments. Moreover, consumers are progressively becoming cognizant regarding problems associated with health, thereby pressing for natural products, which are supposed to promote health and are safe too (Schieber et al., 2001b). The agriculture-food commerce gives rise to large amounts of squanders, both solids and fluid, produced from manufacturing, arrangement, and utilization of food. Additionally, food industries utilize large amounts of water, and the fraction of water may depart as a portion of products (blanching water, olive mill wastewater, etc.). In several situations, cheap source for acquiring bioactive compounds together with high phenolic products is the byproducts from agrarian-food business (Peng et al., 2018). Thus various studies have been investigated to identify the presence of phenols and determine their antioxidant capacity in different byproducts/leftovers of agricultural and food industries which have expanded in recent past and the important byproducts selected have been outlined and evaluated as the source of antioxidants (Balasundram et al., 2006; Shahidi, 2008; Peng et al., 2018).

2.3.10 BY PRODUCT UTILIZATION OF FRUIT AND VEGETABLE INDUSTRY

In recent years, the processing of fruits and vegetables have tremendously increased because of increase in demand and consumption of natural byproducts, so as to live a healthy life (Mullen et al., 2015). Fruit and vegetable refuse which is generated from various food industries and other wholesale markets is enormous. It includes unprocessed fruits and vegetables like tomato, mango, pineapple, jack fruit, orange, banana, etc., which constitute essential component of agrarian wastes. Because of its perishability and environmental hazards, landfill disposal of fruit and vegetable wastes (FVWs) becomes difficult (Pattanaik et al., 2019). It has been reported that in India, the Philippines, China, and United States, approximately 1.81, 6.53, 32.0, and 15.0 mT fruit and vegetable refuse are generated from processing, packing, and distribution, respectively (Wadhwa and Bakshi, 2013). Approximately 5.60 MT is the annual production of fruit and vegetable refuse in India, and presently, the waste generated are dumped on the exterior of cities (Bouallagui et al., 2005). In nature, waste from fruits and vegetables are analogous to food itself.

The large volume of taint discharge is produced by vegetable washing water, which holds soil and other organic matter. Processes like blanching of vegetables and other preparation of food produce wastewater in concentrated form (Hansen and Cheong, 2019). Cannery refuse water is analogous to domestic kitchen refuse. The refuse evolves from trimming, cutting, blanching, and juice processing of fruits and vegetables. The refuse water contains large amounts of mixture of suspended and dissolved solids, the major substance/part being fruit sugars and starch, e.g., the refuse produced from pineapple cannery is 85% to 90% of sugar(sucrose) (Nemerow and Agardy, 1998). By using solid-state fermentation (SSF), the protein content of the F and V processing refuse with ample amount of fermentable carbs can be enhanced by 20–30%. Fruit and vegetable refuse contain enormous amount of fermentable sugar, amongst which apple, oranges, carrots, peas have been effectively used as substratum in fermentation. Also, vinegar, acetic acid, and citric acid have also been produced from the refuse of fruit and vegetable processing (Khedkar and Singh, 2015).

The refuse produced by Sugar beet is composed of 95% of sugar (sucrose) while the remaining is composed of raffinose. This refuse is low in phosphorous and nitrogen. These sugars get leached through bruises and damaged surfaces into the transport and wash-water circuits. These sugars are broken down into short-chain aliphatic carboxylic acid (SCACA) in the circuits, such that the refuse water which requires treatment encompasses solely acids. But in acidic conditions, unpleasant odors from sulfides and VFAs can be produced. In order to maintain the circulating water at neutral pH, adequate amount of lime should be added (Shore et al., 1984). Strawberry which is a seasonal fruit is highly perishable, which can affect its quality and can lead to substantial loss, with destructive effect to economy and environment. Owing to the chemical composition of strawberry, it can be used to acquire newer products for example strawberry puree can be biotransformed, which provides products with enhanced value and health benefits. Glucose in the strawberry puree can be metamorphosed into gluconic acid, in spite of the capacity of *Gluconobacter* sps. to utilize the acid for more biotransformations. Additionally, bioconversion can also be attained without utilizing excess fructose for providing sweetness to the end product. The end product can be mixed with vinegar in variable proportions and can be utilized to produce a wide range of beverages and condiments (Charis, 2020).

Khedkar and Singh (2016) reported the following methods of disposing of the fruit and vegetable refuse are:

- Using landfill;
- Drying the refuse to a moisture content of 10%, so can be used as an animal feed amid out of season;
- Single-cell protein can be produced, if the refuse is biotechnologically processed;
- Refuse from potato and wheat starch manufacturing industries can be fermented to ethanol;
- Peels, membranes, and other components can be processed into juice;
- Fiber pectin, which are high in fiber can be retrieved from lime peels. Hesperidin and eriocitin are present in lemon refuse;
- In citrus fruits and vegetables, flavonoids like hesperidin, naringin, narirutin, and eriocitin are present;
- Citrus refuse can be utilized as a clouding agent due to the presence of a coloring agent.

2.3.10.1 BY PRODUCTS OBTAINED FROM FRUITS

Fruits belong to nourishment items that are cherished by diet and nutrition specialists. Bulks of fruits are loaded with carotenoids, vitamin C and polyphenol mixes (Wijngaard et al., 2009). Diverse phenols and their phenolic content in fruit byproducts are shown in Table 2.1.

2.3.10.1.1 Apple

Apple (*Malus domestica*)is one of the most frequently ingested by mankind (Coman et al., 2019). These are an excellent provenance of $C_{21}H_{24}O_{10}$ and phenolic acids such as (ferulic, p-coumaric, chlorogenic, GA, and caffein), antioxidants, phenols, catechin, epicatechin, quercetin, and naringenin imitatives (Du et al., 2019). Industrial processing manufactures largely juices and cider, frozen, and dried produces, creating tons of byproducts each year (Hesas et al., 2013). Refuse/waste produced by subsequent preparation of apple juice is called apple pomace, which exemplifies up to 30% of indigenous fruit. It primarily includes peels/flesh (95%), stem

TABLE 2.1 Phenolic Content and Compounds in Different Fruit Based Agri-Food Byproducts

SL. No.	Fruit Byproducts		Phenolic Content	Prominent Phenolic Compounds	Functions	References
1.	Apple	Peel	33.42 mg/g (GAE)	Cyanidin glycosides, chlorogenic acid, catechins, caffeic acid, phloridzin, quercetin glycosides, phloretin glycosides, procyanidins	Cardioprotective, antimicrobial, and anticancerous	Panzella et al. (2020); Wijngaard et al. (2009); Wolfe and Liu (2003); Schieber et al. (2003a); Ćetković et al. (2008); Schieber et al. (2003a); Lu and Foo (1998)
		Pomace	4.22–8.67 (CGE)	Phloridzin, chlorogenic acid, quercetin glycosides		
		Seed	2.17 mg/g	Amygdalin, phloridzin		
2.	Mango	Peel	55–110 mg/g (GAE)	Xanthone-C-glycosides, flavonol-O-glycosides	Anti-inflammatory, antioxidative neuroprotective and cardioprotective	Burton-Freeman, Sandhu, and Edirisinghe (2017); Ajila et al. (2007); Schieber et al. (2003b); Puravankara et al. (2000); Soong, and Barlow (2004); Panzella et al. (2020)
		Seed	117 mg/g (GAE)	Gallic acid, ellagic acid, gallates		
3.	Grape	Skin	–	Catechin, epicatechin, gallate, epigallocatechin	Cardioprotective, neuroprotective, anti-inflammatory, antimicrobial, and anticarcinogenic	Souquet et al. (1996); Anastasiadi et al. (2009); Püssa et al. (2006); Yi et al. (2009); Panzella et al. (2020)
		Seed	325.37–811.95 mg/g (GAE)	Dimeric procyanidins B2 and B3, epicatechin-3-ogallate, epicatechin, catechin		
		Stem	367.1–494.2 mg/g (GAE)	Trans-resveratrol, e-viniferin		
		Pomace	107.12–376.71 mg/g (GAE)	Trans-resveratrol, phenolic acids, flavonol glycosides, catechins, anthocyanins		

TABLE 2.1 *(Continued)*

SL. No.	Fruit Byproducts		Phenolic Content	Prominent Phenolic Compounds	Functions	References
4.	Anana	Peel	11 mg/g (GAE)	Dopamine, flavanone glycoside, naringin, rutin	Anti-inflammatory, antiaging, antioxidative effect, effective neurotransmitters	Pazmino-Duran et al. (2001); Coman et al. (2019)
		Bracts		Anthocyanin		
5.	Walnut	Skin (pellicle)	230–490 mg/g (GAE)	Juglone, syringic acid, ellagic acid, condensed, and hydrolyzable tannins	Antioxidative	Labuckas et al. (2008)

Abbreviations: **GAE:** gallic acid equivalents; **CGE:** chlorogenic acid equivalents. Expressed as dry weight.

(1%), and seeds (2–4%) (Bhushan et al., 2008). Freshly pressed remains of apple (apple pomace) should be dried rapidly to hinder the microbial spoilage owing to the high moisture content (75–80%). Apple pomace includes 35–60% dietary fiber (pectins 5–10%, hemicelluloses 4–25%, cellulose 7–40%, lignins 15–25%) (Dhillon et al., 2013) and different materials like tannins, resins, reducing sugars, pigments, that are named as '*extractives*' due to their dissolvability in organic solvent (acetone) or water (Vendruscolo et al., 2008).

Fruitful attempts are being done to utilize apple residue as a source of dietary fiber in bakery goods (Figuerola et al., 2005), meat commodities, (Choi et al., 2016; Yadav et al., 2016) and yogurt (Wang et al., 2019). In another investigation, researchers demonstrated that apple pomace along with oat bran is capable of stabilizing oil-water emulsions (Huc-Mathis et al., 2019). Phloridzin, a chalcone subordinate, is a noted distinctive apple polyphenol which protects the apple plant against several microbes (Veberic et al., 2008). In addition, apple pomaces possess an antidiabetic action by causing interference with the absorption of glucose while obstructing the Na-linked $C_6H_{12}O_6$ carriers (Manzano and Williamson, 2010). Because of its antioxidant character, it can hinder lipid peroxidation, obstruct the bone damage, increase memory and even obstruct the development of cancerous cells (Nair and Rupasinghe, 2014).

2.3.10.1.2 Mango

Mango (*Mangifera indica* L. *anacardiacea*) is one of the foremost tropical fruit in the universe and among the major fruit crops, which is ranked 5th in the overall world production (FAO, 2004). They contain high content of polyphenols ascorbic acid, lupeol, and β-carotene (Sanchez et al., 2013). As mango is a seasonal commodity, different products like puree, canned slices, nectar, pickles, leather, and chutney are processed from about 20% of fruit (Tyug et al., 2010). 13–16% (peels) is the byproducts and seeds (9.50–25%) that are produced during the handling of mangoes (Serna et al., 2015; Torres-León et al., 2017). Peel comprises about 15–20% of total weight of mangoes (Ueda et al., 2000).

A major fraction of mango peels contain 225–725 mg/g (dry wt.) of soluble and insoluble dietary fibers (glucose, galactose, and arabinose being the key neutral sugars), and about 100 mg/g phenolic compounds together with hydrolysable tannins, flavonoids, alkylresorcinols, and proanthocyanidins. Depending on the processing methods, type, and the stage of maturity/ripeness, the qualitative and quantitative composition of phenols also varies accordingly (Jahurul et al., 2015; Schieber, 2017). Also, the mango seeds are abundant in phenols and displays high antioxidant character (Dorta et al., 2014; Torres-Leon et al., 2016). Intriguingly, properties of phenols in mango peel were found to be distinct when compared to the phenols present in the seeds from the similar variety of mango (Dorta et al., 2014), with gallates and GT being the major set in peels, while $C_{14}H_6O_8$ and its prototypes in seeds. This was discovered by another investigation which investigated that the prototypes of $C_8H_8O_5$ and $C_7H_6O_5$ are the key constituents of mango skin (Pierson et al., 2014).

With these interesting properties, various investigations illustrated that the powders from mango peel and its seeds might be added as a promising source for functional food ingredients in foods, like macaroni and biscuits (Ajila et al., 2010; Ashoush and Gadallah, 2011), thus enhancing their potential bioceutical attributes. Mango byproducts can also be utilized for the manufacturing of economical coverings and active packaging substances. Like, addition of mango peel extracts in gelatin-based coatings for active food packaging, which have excellent antioxidant activity and also ameliorates its coating strength (Adilah et al., 2018). In addition to this, the consumable coatings fabricated only of mango peel exhibited excellent perviousness and aquaphobic characteristics (Torres-Leo'n et al., 2018).

Different extracts of mango and individual constituents exhibited properties which proved to safeguard humans and animals against inflammation, oxidation, obesity, heart ailments, diabetes, skin cancers protection to nervous system and gut health (optimistic effects on the intestinal microflora) (Burton-Freeman et al., 2017). Pectin obtained from mango peel has the ability to be utilized in food, cosmetics, and medicinal industries (Geerkens et al., 2015; Sirisakulwat et al., 2010). Pectin played a strong part in avoiding and decreasing the chances for cancers; thus, it has grabbed the attention of the scientific communities (Maxwell et al., 2012).

2.3.10.2 BYPRODUCTS OBTAINED FROM VEGETABLES

The different phenolic compounds and its phenolic content in various vegetable byproducts are shown in Table 2.2.

TABLE 2.2 Phenolic Content and Compounds in Different Vegetable Based Agri-Food Byproducts

SL. No.	Vegetable Byproducts		Phenolic Content	Prominent Phenolic Compounds	Functions	References
1.	Red beet pomace	–	87–151 mg/g (GAE)	l-Tryptophan, p-coumaric acid, ferulic acid	Effective against bacteria, cancer, hypertension hepato organs, antioxidant, anti-inflammatory, antianemia, and detoxifier	Peschel et al. (2006); Chhikara et al. (2019)
2.	Olive	Leaves	–	Oleuropein, hydroxytyrosol, luteolin-7-glucoside, apigenin-7-glucoside, verbascoside	Antioxidative, anti-inflammatory, anti-atherogenic. Effective against cancer, microbes, and viruses. Hypolipidemic and hypoglycemic, provides assistance to patients suffering from Alzheimer's	Benavente-Garcia et al. (2000); Ramos Cormenzana et al. (1996); Al-Azzawie and Alhamdani (2006); Bisignano et al. (1999); Bulotta et al. (2014); Cordero, Garcı´a-Escudero, Avila, Gargini, and Garcı´a-Escudero (2018); Omar (2010); Soler-Rivas, Espi_n, and Wichers (2000); Visioli, Bellosta, and Galli (1998)
		Mill wastewater	–	Hydroxytyrosol, tyrosol, oleuropein, dimethyloleuropein, verbascoside, catechol, hydroxycinnamic acids, 4-hydroxybenzoic acid		
3.	Carrot	Peel	13.8 mg/g(GAE), 9.79 mg/g (GAE)	Chlorogenic and caffeic acid	Healing wounds, antiobesity, antiviral, antimutagenic, anti-inflammatory, and cardioprotective	Chantaro et al. (2008); Zhang and Hamauzu (2004); Panzella et al. (2020)

TABLE 2.1 *(Continued)*

SL. No.	Vegetable Byproducts		Phenolic Content	Prominent Phenolic Compounds	Functions	References
4.	Tomato	Skin	0.29 mg/g (GAE)	Quercetin, kaempferol, rutin, phenolic acids, naringenin	Cardioprotective and anticancerous	Toor and Savage (2005); Coman et al. (2019)
5.	Red onion	Dry peel	384.7 mg/g (GAE)	Quercetin, quercetin glycoside	Anticarcinogenic, antidiabetic, anticholestrol, antiinflammtory, antiosteoporosis, antifungal, antimicrobial	Singh et al. (2009); Lee et al. (2008); Sharma et al. (2016)

Abbreviations: **GAE:** gallic acid equivalents; CGE: chlorogenic acid equivalents. Expressed as dry weight.

2.3.10.2.1 Carrot

The carrot (*Daucus carota*) belongs to the Apiaceae family (Umbelliferae) and is the most valued root vegetable, which can be used in its unprocessed or cooked or in the form of juices (Coman et al., 2019). Carrots have rich antioxidant profile and are an excellent source of anthocyanidin, carotenoid, and fiber content. Additionally, carrots contain antioxidant, anti-cancer, and anti-inflammatory compounds and support cardiovascular health and reduce the development of degenerative disease (Clementz et al., 2019). Carrots can also be used in grated form in carrot puddings and carrot cakes. Mini-carrots or baby carrots (after peeling) are available in many supermarkets, consumed as a ready-to-eat snack. Due to its impressive health benefits and various nutritional values, carrot juice is the most widely consumed health drink. During all these processes, peels, and pomace are derived in large quantities which are a good source of different valuable constituents, principally bioactive substances with antioxidant properties (Chantaro et al., 2008). Carrot peels have also been characterized as a good reserve of various phenolic substances (Chantaro et al., 2008). Besides hydroxycinnamic acids and its prototypes such as 4-p-coumaroylquinic acid, caffeic acid, 3-caffeoylquinic acid, 3,5-dicaffeoylquinic acid, 3,4-dicaffeoylquinic acid, $C_{16}H_{18}O_9$ is the most plentiful phenolic acid found in carrots (Zhang and Hamauzu, 2004). Harding and Heale (1980) identified 6-methoxymellein and isocoumarins 6-hydroxymellein in carrots. In a comparative research done by Chantaro et al. (2008), the quantity of phenolic substances were analyzed in carrot peels under varied working environments and the total phenolic levels **found** were around 1,380 milligrams (GAE)/100-gram dry wt. Pectin is an essential component of carrot skin, a soluble dietary fiber composed of D-galacturonic acid, L-arabinose, L-rhamnose, and D-galactose units and as well as other monosaccharides. Pectin (E440) can be used as an addon or an accompaniment in functional foods such as jellies, jams, frozen foods or low-calorie foods. Pectin, a structural acidic heteropolysaccharide, displaying thickening and stabilizing properties offers great potential for replacement of hydrocolloids (Encalada et al., 2019). Other beneficial components of carrot peel are carotenoids (α- and β-carotene) (Clementz et al., 2019).

Carrot enriched cookies with fiber were found to have enhanced the color attributes and showed increased resistance to water absorption, more

Food Processing Byproducts 51

stability to dough and its growth and less softening (Turksoy and Ozkaya, 2011). Results also showed that fibers possessed high antioxidant and prebiotic characteristics (Liu et al., 2019).

Carrot was the primary vegetable to be supported as secure to be used within the biopharmaceutical sector. In addition of having a good nutritional profile, the consumption of carrot or carrot byproducts or carrot-enriched products have a favorable effect on hypertension, constipation, cardiac ailments, cancer, infections, atherosclerosis, bronchial asthma, control blood sugar level and prevent muscular degeneration. Carrot peels are an excellent source of polyphenols that protect cells from damaging free radicals, have antitumor and anti-mutagenic actions (Sharma et al., 2012). Enzymes and constituents are being obtained from carrots, which are effective against different diseases, e.g., human deoxyribonuclease I in cystic fibrosis, glucocer-ebrosidase in Gaucher's disease, α-galactosidase-A in Fabry disease and also the manufacturing of various derivatives of vaccine is achiev-able in carrot tissues (Mendoza and Olea, 2015). The constituents of carrot retinol and its metabolite, $C_{20}H_{28}O_2$, enact a critical function in the development of fetus, reproduction, functioning of lungs, and immunity (antioxidant effects and anti-inflammatory) (Hufnagl and Jensen-Jarolim, 2019).

2.3.10.2.2 Onion

Onion (Alliumcepa L.) is the most significant vegetable crops, found in countless recipes and these days accessible in fresh, chopped, frozen, powdered, pickled, canned, and dehydrated form (Shahidi et al., 2019). The onion byproducts primarily consist of outer papery brown, red, or yellow skins produced during mechanical peeling; upper part of roots and lower part of bulb, outer hefty scales, and miniatured, diseased, distorted, or disfigured onion bulbs (Sharma et al., 2016). The onion scales which are an excellent source of phenolics and organosulfur constituents are thick and white (Khiari et al., 2009). Onion refuse is also a great reserve of dietary fiber, fructans, flavonoids, and alkenyl cysteine sulfoxides (Table 2.2; Gregorio et al., 2014), which are useful to the rehabilitation of humans (Nile et al., 2017; Tram et al., 2005). In onion, a total of 16 different flavonols are identified, which include aglycones and glycosylated prototypes of quercetin (Ko et al., 2015). The studies confirmed these flavonols compounds as quercetin 3-glucoside, quercetin 3,4-diglucoside, isorhamnetin 4-glucoside, isorhamnetin 3,4-diglucoside, quercetin 4-glucoside, and forming about 90% of the onion bulb, internal, and external scales (Benitez et al., 2011). Dry peel (outer layer) is the major byproduct of onion which is rich in quercetin glycoside, and their oxidative product. These compounds are potent antioxidants, providing defense against cellular damage and are highly reactive against oxidative stress-induced damage (Prakash et al., 2007). Besides quercetin substances, dried onion peels are a good source of ferulic, kaempferol, protocatechuic acids and Gallic (Singh et al., 2009). A comparative study by Velioglu et al. (1998) studied that the total phenolic content in red onion scale was 105.50 gram/ kilogram dry wt. In another study, it was found as 385 milligrams (GAE)/g and 165 milligrams QE/gram respectively in $C_4H_8O_2$ proportion red dried onion skin by analyzing various solvent extracts (dichloromethane extract and diethyl ether, ethyl acetate, and n-butanol fractions of ethanolic extract). In addition, ethyl acetate fraction possessed higher antioxidant, free radical scavenging and reducing power activities in comparison to other proportions (Singh et al., 2009). The extract obtained from onion leaf was established to be a steady natural colorant and offers a promising substitute to artificial colorants (Mourtzinos et al., 2018).

Quercetin was shown to be potentially effective against allergies, cancer, asthma, coronary heart disease, arthritis, osteoporosis neurodegenerative disorders, gout, and diabetes (Curti et al., 2017; Nile et al., 2017). Onion waste/byproducts is abundant in enormous beneficial bioactive phytochemicals, which have potential utilization in cosmetic and pharmaceutical industries (Nile et al., 2018). The health-associated impacts of onions are co-related with anti-diabetic, anti-cholesterol, anti-carcinogenic, anti-inflammatory, antifungal, anti-microbial, antioxidant, anti-osteoporosis, and hypotensive activities (Lee et al., 2008; Sharma et al., 2016).

2.3.10.3 BYPRODUCTS OBTAINED FROM MEAT INDUSTRY

In the meat industry, the maximum waste is generated through slaughtering. By carnaging and processing of animal, only one third is meat while the rest encompasses byproducts and refuse, and the wastes thus obtained need to be processed effectively and then utilized efficiently. Each partition of a slaughtered animal is byproduct apart from dressed carcass (also called offals or fifth quarter) (Irshad et al., 2015). Wastes in the meat industry include skin, tendons, bones, and the parts of the GIT, blood, and internal organs (Kulkarni and Devatkal, 2015). Wastes generated from various type of animal are different (Fleischtechnologie, 1996; Grosse, 1984). For the profitability, it is necessary that meat byproducts should be utilized efficiently by meat industries (Jayathilakan et al., 2012. Efficient utilization of byproducts encompasses a straight influence on the financial system and polluting the surroundings of the nation, the previous being an asset and the latter being a responsibility (Irshad et al., 2015). On the basis, whether animal wastes can be used as food or not, they are classified as edible byproducts or inedible byproducts. Edible byproducts are those that can be consumed as food by humans and comprise the brain, liver, spleen, kidney, heart, tongue, intestine, etc. Contrarily byproducts that cannot be consumed as food by humans are called inedible byproducts, e.g., hides, skins, hair ear, hoofs, snout, gallbladder, fetus, horns, bristles, etc. It also includes organs of dead animal or condemned meat (Sharma, 2011). Eatable meat byproducts contain various vital nutrients like liver and kidney encloses a higher content of carbohydrate than other meat materials (Devatkal et al., 2004). The highest fat and lowest moisture content is present in pork tail as compared to other meat byproducts. The protein content of the liver, tail, ears, and feet of cattle is approximately the same as lean meat tissue, but a sufficient quantity of collagen is found in the ears and feet (Unsal and Aktas, 2003). Byproducts possess a large amount of connective tissue, so they have different composition of amino acids than lean tissues. Therefore, byproducts like feet, lungs, ears, stomach, and tripe holds a large portion of glycine, $C_5H_9NO_3$ and proline, and a lesser level of tryptophan and tyrosine. Also, organ meat has generally more vitamin content than lean meat tissue (Jayathilakan et al., 2012). An important edible byproduct is animal blood that has a high content of protein and heme iron (Wan et al., 2002). Various products like blood pudding, blood sausages, biscuits, and bread are made from animal blood in Europe. In Asia, it is utilized in blood cake, blood curd and blood pudding (Ghost, 2001). It is also utilized as non-food items like fertilizer, binders, and feedstuffs. Plasma, part of blood is of significant use due to its functional properties and lack of color (Jayathilakan et al., 2012). Apart from this, it has a brilliant foam-forming capability (Del Hoyo et al., 2008), and can be used as best replacers for egg whites in bakery industry (Ghost, 2001). Since prehistoric times, animal byproduct (hides)has been used as shelters, clothing, and containers by humans. Gelatin can also be obtained from the skin of animals and hides and can be used in food (Choa et al. 2005). It can be incorporated in various foods, further as forming a key constituent in jellies and aspic (Jamilah and Harvinder, 2002). In meat products, collagen from hides and skins can be utilized as an emulsifier for it can bind huge amounts of fat and thus can be utilized as filler or additive for meat products. It can also be used to make the collagen sausage. A product prepared from extracted collagen can act as stimulant for blood clotting in surgery (Jayathilakan et al., 2012). In various countries animal intestines are used as food after boiling them. They are conjointly utilized in pet food or for meat meal, tallow or fertilizer. Sausage casings are the major significant product of intestines (Bhaskar et al., 2007). Animal products which are good source of cholesterol are brains, nervous systems and spinal cords and also used for production of cholecalciferol. Cholesterol is used in the cosmetic industry as an emulsifier (Ejike and Emmanuel, 2009). Heparin, derived from the liver, as well as from lungs and the lining of small intestines can be utilized as an anticoagulant to extend the time for blood clotting (Jayathilakan et al., 2012).

Squander products from the poultry processing and egg-producing industries need to be dealt with efficiently because the development of these industries mostly rely on their refuse management. Shih (1993) reported that an efficient TADS that changes animal dung to methane for an energy source was done by *Bacillus licheniformis*, which is a feather degrading bacteria that can ferment and change feathers to their lysates, which is a digestible protein source and can be used as feed (Jayathilakan et al., 2012).

Fish refuse is an excellent source of proteins, minerals, and fat. Khan et al. (2003) investigated that the FPH can be produced by enzymatic treatment of refuse from 5 marine fish scrap species (white croaker, flying fish, horse mackerel, Sardine, chub mackerel). Chitosan can be prepared from shrimp and crab shell which has revealed a broad range of applications from cosmetic to pharmaceutical industries (Arvanitoyannis and Kassaveti, 2008). Chitosan also has many applications in food preservation such as, direct application of chitosan film, direct addition of chitosan into food, or coatings onto food surfaces, addition of chitosan sachets into packages, and use of packaging materials incorporated with chitosan. The fish oil produced from byproducts of matjes (salted) and its stability was studied to estimate the quality attributes (Aidos et al., 2001).

2.3.10.4 BYPRODUCTS OBTAINED FROM DAIRY INDUSTRY

Dairy industry continually produces a large volumes of proteinaceous wastes, majority of which being generated as refuse water from cleaning of milk tanks and other dairy handling processes (Arvanitoyannis and Kassaveti, 2008). Wastewater of dairy usually includes high conc of dissolved organic constituents, like fat, lactose, minerals, and whey protein. Dairy industry produces a number of byproducts such as buttermilk, whey, skim milk, and GR and derived byproducts like caseins, caseinates, lactose, WP, etc. (Rafiq and Rafiq, 2019). Whey is the foremost byproduct of the dairy industry that is generated while cheese making. Milk whey is rich in fats, carbohydrates, soluble vitamins, minerals also proteins. About 75% of lactose is present in total whey solids (De Jesus et al., 2015). Various attempts have been made worldwide to use these byproducts prior to their lofty nutritional value (Rafiq and Rafiq, 2019).

Recently, whey achieved attention as a food constituent, approaching into use as a technological medium mostly through whey proteins, accomplishing an exquisite merge between its nutritional and functional characteristics, which has relevance both in food and health. Selective porous membranes using isolation and fractionation processes are being used to separate whey components. They can be used as supplements in the agri-food industry, such as athletic drinks; still, 40% of whey remains unutilized (Bugnicourt et al., 2010; Onwulata and Huth, 2009). Direct fermentation of whey can be achieved using lactose-consuming microorganisms while aiming at bioconversion, which will result in the outcome of various products (Panesar et al., 2013). Polyhydroxy butyrate (PHB)can be produced from co-metabolization of $C_6H_{12}O_6$ and galactose from lactose-hydrolyzed by *Pseudomonas hydrogenovora.* (Koller et al., 2008). This bio-plastic has a huge impact in the medical field and is ecologically safe. Furthermore, it can facilitate the delivery of drugs to the body by the production of nanoparticles (Pandian et al., 2010). By using *Xanthomonas campestris*, heteropolysaccharide like xanthan gum can be produced from whey (Mesomo et al., 2009). WPI and WPC are broadly used in food applications due to their higher protein contents, low energy, fat, and Na contents, absence of microbes and deadly components; biocompatibility and GRAS status, readily available, and inexpensive products (Rafiq and Rafiq, 2019). Caseins might be merged with different substances in different proportions so as to manufacture a composite (Aberg et al., 2004). Caseins have also been utilized to make adhesives and glues for diverse purposes (Audic et al., 2003).

GR has been utilized in food firms for manufacturing of sweets, bakery products, and as a flavor enhancer (Tamine, 2009). Various products can be prepared by blending ghee residue with SMP like chocolate, khoa, and sugar, burfi (Verma and De, 1978). After mixing with other components, it could also be used for manufacturing of toffees, coconut burfi, pinni, candies, etc. The healthful byproducts must be utilized as a food additive in a number of foods, food spreads, soups, etc. (Galhotra et al., 1993). GR is an excellent supply of phospholipids from which phospholipids can be retrieved and added to ghee (Rafiq and Rafiq, 2019). Buttermilk concentrate is also a rich source of phospholipids, which has been used in processed cheese spread to enhance its organoleptic, rheological, and functional characteristics (Sayed et al., 2010). Buttermilk has been added to lessen-fat cheese up to 40% and was found to develop the sensory score in contrast with control (Pinto et al., 2014). Thus, these byproducts are useful in food, fuel health, pharmaceuticals, etc.

2.3.10.5 ENZYMES OBTAINED FROM DIFFERENT TYPES OF WASTE

Under mild conditions of a certain chemical reaction, enzymes which are proteins act as vigorous and localized catalysts. These bio-molecules have important utilization in various economic fields such as food, cosmetics, pharmaceuticals, textiles, chemicals, and fuels (Sharma et al., 2016). They are selected as chemical catalysis in many cases because of their higher degree of specificity for substrates and other vigorous operational parameters. Due to the high raw material and processing cost commercial enzymes are mainly exorbitant. These increasing costs can be lowered by producing enzymes using agro-food waste as the substrate (Ravindran and Jaiswal, 2016). The variable composition of agro-food wastes, which are utilized as a basic/raw material helps in the multiplication of microorganisms as a consequence of ebullition, which gave rise to various beneficial enzymes. These substrates resulted in the enhancement of the fungal growth rate which ended in the metamorphosis of lignocellulosic substrate into lesser arduous ones by degeneration of various enzymes (Sadh et al., 2018). Furthermore, these substrates can yield a great variety of enzymes shown in Table 2.3. Some appropriate examples include xylanases, amylases, proteases, invertase, peroxidases, polyphenol oxidases, pectinolytic enzymes and tannases (Bharathiraja et al., 2017; Panda et al., 2016; Sharma et al., 2016). The manufacturing of enzymes is categorized into two main groups: Submerged fermentation (SmF), and solid-state fermentation (SSF) (Kapoor et al., 2016). The mode of operation in case of SmF is either batch or semi-continuous type in which the microorganisms are grown in excess of free-flowing water. The insoluble substrates are either submerged or suspended while as the soluble substrates are diffused in a liquid state. This design authorizes a fine command over different reaction conditions. Even though the potency is less, there is the possibility of strong effect of important obstructing/ impeding substances, and the provision for energy is comparably high. In comparison SSF involves the growth of microorganisms devoid of free liquid phase on solid, moist substrates. This design is feasible for the development of fungi and yeast, rather than bacteria (Viniegra-Gonzalez et al., 2003). SSF is advantageous over SmF as it requires less energy, low capital cost, and easy ebullition medium; produces less affluent and has higher efficiency (Couto et al., 2006). Furthermore, it becomes more attractive because of the less capital cost linked with the down-streaming process and an easy control over the bacterial contamination. SSF yields effective environmental conditions for its development and production of enzymes analogous to the natural environment of microorganisms (Thomas et al., 2013). Although very fewer reports are available in literature on SSF bioreactor design. Using SSF, it becomes difficult to control pH, aeration, oxygen transfer temperature, and moisture content and hence cannot be utilized for large scale manufacturing of enzymes (Couto et al., 2006; Wang et al., 2008).

58   Integrated Waste Management Approaches for Food and Agricultural Byproducts

TABLE 2.3   Enzymes Produced by Microorganisms Using Agri-Food Waste

SL. No.	Enzyme	Substrate	Microorganisms	References
1.	Protease	Pieces of bread refuse	*Aspergillus awamori, Aspergillus oryzae*	Melikoglu et al. (2013, 2015)
		Date waste	*Bacillus* species. 2–5	Khosravi et al. (2008)
		Potato waste	*Saccharomyces cerevisiae*	Afify et al. (2011)
2.	Amylase	Refuse from kitchen	*Chryseobacterium Bacillus* sp.	Hasan et al. (2017)
		Banana skin	*A. Niger NCIM 616*	Krishna et al. (2012
		Potato peel	*Bacillus subtilis* *Bacillus licheniformis* *Bacillus firmus* CAS 7	Kiran et al. (2014)
3.	Xylanase	Peel from citrus waste	*A. Niger*	Ahmed et al. (2016)
		Pomace of grape	*A. awamori*	Botella et al. (2007)
4.	Lipase	Oil cake (groundnut)	*C. rugosa*	Rekha et al. (2012)
		Banana, melon, watermelon refuse	*B. coagulans*	Alkan et al. (2009)
		Refuse from cooking olive oil	*Aspergillus* and *Penicillium* strains	Papanikolaou et al., (2011)
5.	Cellulase	Soybean hulls	*A. Niger* NRRL3	Julia et al. (2016); Bansal et al. (2012)
		Kitchen refuse pretreated with Alkali	*A. Niger* NS-2	

2.3.10.5.1 Amylase

Amylases are enzymes that break down starchy carbohydrate units like glucose, maltotriose, and maltose (Sindhu et al., 2019). The two major phases in the amylase family are glucoamylase (GA) (EC 3.2.1.3) and α-amylase (EC 3.2.1.1). Hydrolysis of starch into $C_{12}H_{22}O_{11}$, sugar, and maltotriose by removing 1,4-α-D-glucosidic interactions between sugar-related units within the lines of amylose chain is caused by α-amylase (Pandey et al., 2000) while as GA hydrolyzes result in a reduction of amylose and amylopectin in glucose (Anto et al., 2006). These amylases

Food Processing Byproducts 59

receive a good range of applications, performing the production processes of plant extracts, bread, beer cloth, paper, gin, and medicine (Sahnoun et al., 2015). The carbon and carbon source of carbon is often used in the assembly of amylases. The use of additional agricultural residues such as food and kitchen food is another source of less expensive amylase production (Sindhu et al., 2019). Krishna et al. (2012) used an amylase assembly page by *Aspergillus Niger* NCIM 616. Research has shown strong resilience as a promising strategy compared to lower immersion. Other studies included banana refuse (pre-processed stem with banana fruit by automatic extraction at 0, 1, 2°C for 60 minutes) as a meeting point for high alpha-amylase particles (minimal 5,345,000 U/ milligram min) using *Bacillus subtilus* utilized CBTK 106 (Krishna et al., 1996). Amylase enzyme was processed from different species of fungus and bacteria with varied aspects such as thermos-solid, halotolerant, tolerance to extreme low temperatures and solid-alkali (Prakash et al., 2019). Coffee refuse was used as the only source of carbon for the concoction of α-amylase in the SSF using Neurosporacrassa CFR 308. At 28°C temperature, 60% moisture content and pH 4.6, an α-amylase activity of 4,324 U/gram dry substrate was employed using 1.0 mm particle size, spores/g substrate dry matter. Coffee contamination and activity of α-amylase 6,342 U/gram dry substrate got enhanced with the pre-steam process (Murthy et al., 2009).

2.3.10.5.2 Protease

Proteases are enzymes which are important hydrolytic enzymes that carry out proteolysis by incorporating hydrolysis in peptide bond by binding amino acids together with the polypeptide chains. These enzymes always need to move forward, because of their active roles in cellular metabolism, use in food, medicine, detergents, dairy, leather industries and silver refining from used X-ray films (Ridhi and Sarawathi, 2014; Jisha et al., 2014).

Melikoglu et al., (2015) examined the synthesis of protease by Aspergillus awamori using pieces of bread crumbs as substrate during the full iron reaction. The maximum activity of protease was 80.3 U/gram bread when airflow was maintained at 1.50 VVM. Studies show the potency of bread contaminants as the potential for proteases production. German et al. (2003) used the soybean cake as a substrate to support the protease assembly of Penicillium species. The impure enzyme retains its function (50–60%) within the economics. Khosravi et al. (2008) employed the use of newly developed Bacillus sp. which was alkalophilic in nature in the pre-drug SMF with no prior treatment. At pH 10 and a temp of 37°C high protease (57,420 APU/mL) was found, which declared the enzyme to be highly curable signifying its use in industries. Afify et al. (2011) researched the assemblage of proteases from potato refuse using S. cerevisiae. For plant development, they also studied the use of solid waste as biofertilizer. At a primary pH of 6.0 and temp of 20°C for 72 h simple enzyme activity (360 U/milligram) was attained using a Mediation medium carrying 15 g potato refuse. Protease has also been produced using substrates like pomace of tomato and seed cake from Jatropha (Belmessikh et al., 2013; Veerabhadrappa et al., 2014).

2.3.10.5.3 Xylanase

Xylanases (E. C. 3.2.1.8, 1.4-beta-xylanohydrolase) are the enzymes that disintegrate xylan, and are also an important component of plant polysaccharides. Xylan is an intricate polysaccharide, composed of xylose-residue, with every part connected by a beta-1,4-glycosidic linkage. D-arabinofuranoside and D-glucuronic acid or can be associated with the Xylan spine (Harris and Ramalingam, 2010). These can be used in food, animal feed, biomedical, and bioethanol industries (Harris and Ramalingam, 2010; Das et al., 2012). The grape refuse is the remaining portion after the juice has been drawn from the grapefruit. This is not a good animal feed as it contains low food content and high amounts of phenolic substances. Dumping of grapevine pomace ushers to grave hazards to the habitat. Therefore, it appears favorable to use this value addition. Because of high phenolic compounds grape pomace cannot be used for various fertilizer programs because it prevents the germination of seeds. Botella et al. (2007) tested the ability of grapevine pomace creation/ manufacturing of xylanases by *Aspergillus awamori*. Studies have shown that the addition of a supplementary source of C and the primary m.c of grape pomace enacts an essential part in the making of enzymes (Sindhu et al., 2019). Xylanase production from *Aspergillus Niger* and *Trichoderma viride* using diverse agricultural wastes was examined by Soliman et al. (2012). They proclaimed that after 2 days of implantation at a temp of 35°C and pH 5.5, barely bran was the absolute place for the maximum enzyme

Food Processing Byproducts 61

production (42.5 U/gram substrate) from *Trichoderma viride*. Seyis et al. (2005) used apple pomace; peel from melon and shell of hazelnut in addition to xylanase a byproduct of Trichoderma harzianum 1073 D3, reached a maximal act of 26.5 U/mg.

2.3.10.5.4 Lipase

Lipases receive different types of requests due to the type of response you can issue. In the biological system, lipases are responsible for the secretion and binding of lipids within cells. According to the reaction rate, lipases incorporate other exciting technological processes, including hydrolysis, dispersing, alcohol extraction, esterification, acidolysis, and aminolysis.

It can turn to hydrolyze oils into glycerol and acids in water-lipid expression and is able to affect the adaptive response to a media which is not liquid. All these factors drives lipases as the most extensively desired enzyme in managements such as bakery, chemical biodiesel production, biosynthesis production, oil, milk, ports, and saturated fatty acids treatments (Aravindan et al., 2007). Phospholipases belong to the class of lipases. These lipases bind hydrolysis of phosphodiester bond of glycerophospholipids and one or several esters, which exhibits changes in the phospholipid production, which can be used for the conversion/making of newer phospholipids in other operations such as oil refining, health, processing of food, milk, and industrial industries (Ramrakhiani and Chand, 2011). The efficacy and the production of lipase enzyme were performed using oil cakes as substrates from agro-food waste by Oliveira et al. (2017). A study investigated *A. Niger* as a manufacturer of synthesizing lipase enzyme using substrate as peanut cake and 1.0% (w/w) between 80, 0.35% (w/w) ammonium sulfate, and 0.40% (w/w) sodium phosphate (Salihu et al., 2016). Enzyme output of 49.37 U/g was found. Besides, employing the use of shea nut cake, the production of lipase reduces the amount of tannins and saponins, making it suitable for using it as feed for animals (Ravirandran et al., 2018).

2.4 CONCLUSION

Prior to urbanization and modernization, the quantity of wastes from agriculture and agro-industries is increasing at an alarming rate. However, the wastes generated are of paramount significance due to the presence of nutritional substances. The utilization of agricultural refuse can be harnessed to deliver distinct cost-effective benefits to the country and thus improve the quality of the environment. Through the application of agricultural waste management systems (AWMS) such as 3R's, WM can be properly achieved. Various conventional food WM technologies are carried out to discard the waste or to utilize them in a better way. Byproducts can be utilized to produce edible oils, vitamins, enzymes, pigments, livestock feed, alcohols, dyes, and biofuels as they are rich in nutrients and other bioactive substances. Thus it becomes important to look for methods and other in-plant processes, where refuse and its byproducts can be recuperated and improved to a higher product value. Hence the byproducts necessitate to be regarded as a source of economy with capacity for utilization in value addition and not as refuse.

KEYWORDS

- **agro-industrial waste**
- **biodiesel**
- **food waste**
- **pyrolysis**
- **toxic waste**

REFERENCES

Abbasi, T., Tauseef, S. M., & Abbasi, S. A., (2012). A brief history of anaerobic digestion and "biogas." In: *Biogas Energy* (pp. 11–23). Springer, New York, NY.

Aberg, C. M., Chen, T., Olumide, A., Raghavan, S. R., & Payne, G. F., (2004). Enzymatic grafting of peptides from casein hydrolysate to chitosan. Potential for value-added byproducts from food-processing wastes. *Journal of Agricultural and Food Chemistry, 52*(4), 788–793.

Abou, H. S. D., & Sawan, O. M., (2010). The utilization of agricultural waste as one of the environmental issues in Egypt (a case study). *Journal of Applied Sciences Research, 6*(8), 1116–1124.

Adilah, A. N., Jamilah, B., Noranizan, M. A., & Hanani, Z. N., (2018). Utilization of mango peel extracts on the biodegradable films for active packaging. *Food Packaging and Shelf Life, 16*, 1–7.

Food Processing Byproducts

Afify, M. M., El-Ghany, T. M. A., & Alawlaqi, M. M., (2011). Microbial utilization of potato wastes for protease production and their using as biofertilizer. *Australian Journal of Basic and Applied Sciences, 5*(7), 308–315.

Aidos, I., Van, D. P. A., Boom, R. M., & Luten, J. B., (2001). Upgrading of maatjes herring byproducts: Production of crude fish oil. *Journal of Agricultural and Food Chemistry, 49*(8), 3697–3704.

Ajila, C. M., Aalami, M., Leelavathi, K., & Rao, U. P., (2010). Mango peel powder: A potential source of antioxidants and dietary fiber in macaroni preparations. *Innovative Food Science & Emerging Technologies, 11*(1), 219–224.

Ajila, C. M., Leelavathi, K. U. J. S., & Rao, U. P., (2008). Improvement of dietary fiber content and antioxidant properties in soft dough biscuits with the incorporation of mango peel powder. *Journal of Cereal Science, 48*(2), 319–326.

Al-Azzawie, H. F., & Alhamdani, M. S. S., (2006). Hypoglycemic and antioxidant effect of oleuropein in alloxan-diabetic rabbits. *Life Sciences, 78*(12), 1371–1377.

Al-Hamamre, Z., (2011). *Biodiesel Production from Olive Cake Oil.* University of Jordan, Jordan.

Alexandratos, N., & Bruinsma, J. (2012). World Agriculture towards 2030/2050: The 2012 Revision. ESA Working Paper No. 12-03. FAO, Rome.

Alkan, H., Baysal, Z., Uyar, F., & Dogru, M., (2007). Production of lipase by a newly isolated *Bacillus coagulans* under solid-state fermentation using melon wastes. *Applied Biochemistry and Biotechnology, 136*(2), 183–192.

Anastasiadi, M., Chorianopoulos, N. G., Nychas, G. J. E., & Haroutounian, S. A., (2009). Antilisterial activities of polyphenol-rich extracts of grapes and vinification byproducts. *Journal of Agricultural and Food Chemistry, 57*(2), 457–463.

Anto, H., Trivedi, U. B., & Patel, K. C., (2006). Glucoamylase production by solid-state fermentation using rice flake manufacturing waste products as substrate. *Bioresource Technology, 97*(10), 1161–1166.

Aravindan, R., Anbumathi, P., & Viruthagiri, T., (2007). *Lipase Applications in Food Industry.*

Aruna, G., & Baskaran, V., (2010). Comparative study on the levels of carotenoids lutein, zeaxanthin, and β-carotene in Indian spices of nutritional and medicinal importance. *Food Chemistry, 123*(2), 404–409.

Arvanitoyannis, I. S., & Kassaveti, A., (2008). Dairy waste management: Treatment methods and potential uses of treated waste. *Waste Management for the Food Industries,* 801–859.

Arvanitoyannis, I. S., & Kassaveti, A., (2008). Fish industry waste: Treatments, environmental impacts, current, and potential uses. *International Journal of Food Science & Technology, 43*(4), 726–745.

Ashoush, I. S., & Gadallah, M. G. E., (2011). Utilization of mango peels and seed kernels powders as sources of phytochemicals in biscuit. *World J. Dairy Food Sci., 6*(1), 35–42.

Audic, J. L., Chaufer, B., & Daufin, G., (2003). Non-food applications of milk components and dairy co-products: A review. *Le Lait, 83*(6), 417–438.

Balasundram, N., Sundram, K., & Samman, S., (2006). Phenolic compounds in plants and agri-industrial by-products: Antioxidant activity, occurrence, and potential uses. *Food Chemistry, 99*(1), 191–203.

Baldisserotto, A., Malisardi, G., Scalambra, E., Andreotti, E., Romagnoli, C., Vicentini, C. B., & Vertuani, S., (2012). Synthesis, antioxidant, and antimicrobial activity of a new phloridzin derivative for dermo-cosmetic applications. *Molecules, 17*(11), 13275–13289.

Bani, O., Parinduri, S. Z. D. M., & Ningsih, P. R. W., (2018). Biodiesel production from rice bran oil by transesterification using heterogeneous catalyst natural zeolite modified with K_2CO_3. In: *Materials Science and Engineering Conference Series* (Vol. 309, No. 1, p. 012107).

Bansal, N., Tewari, R., Soni, R., & Soni, S. K., (2012). Production of cellulases from *Aspergillus Niger* NS-2 in solid-state fermentation on agricultural and kitchen waste residues. *Waste Management, 32*(7), 1341–1346.

Becidan, M., Skreiberg, Ø., & Hustad, J. E., (2007). NO x and N_2O precursors (NH_3 and HCN) in pyrolysis of biomass residues. *Energy & Fuels, 21*(2), 1173–1180.

Belmessikh, A., Boukhalfa, H., Mechakra-Maza, A., Gheribi-Aoulmi, Z., & Amrane, A., (2013). Statistical optimization of culture medium for neutral protease production by *Aspergillus Niger*. Comparative study between solid and submerged fermentations on tomato pomace. *Journal of the Taiwan Institute of Chemical Engineers, 44*(3), 377–385.

Benavente-Garcia, O., Castillo, J., Marin, F. R., Ortuno, A., & Del, R. J. A., (1997). Uses and properties of citrus flavonoids. *Journal of Agricultural and Food Chemistry, 45*, 4505–4515.

Benavente-Garcia, O., Castillo, J., Lorente, J., Ortuño, A. D. R. J., & Del, R. J. A., (2000). Antioxidant activity of phenolics extracted from *Olea europaea* L. leaves. *Food Chemistry, 68*(4), 457–462.

Benítez, V., Mollá, E., Martín-Cabrejas, M. A., Aguilera, Y., López-Andréu, F. J., Cools, K., & Esteban, R. M., (2011). Characterization of industrial onion wastes (*Allium cepa* L.): Dietary fibre and bioactive compounds. *Plant Foods for Human Nutrition, 66*(1), 48–57.

Berardini, N., Knödler, M., Schieber, A., & Carle, R., (2005). Utilization of mango peels as a source of pectin and polyphenolics. *Innovative Food Science & Emerging Technologies, 6*(4), 442–452.

Bhada, P., (2007). *Feasibility Analysis of Waste-to-Energy as a Key Component of Integrated Solid Waste Management in Mumbai, India*. Master of Science in Earth Resources Engineering, Columbia University.

Bharathiraja, S., Suriya, J., Krishnan, M., Manivasagan, P., & Kim, S. K., (2017). Production of enzymes from agricultural wastes and their potential industrial applications. In: *Advances in Food and Nutrition Research* (Vol. 80, pp. 125–148). Academic Press.

Bhaskar, N., Modi, V. K., Govindaraju, K., Radha, C., & Lalitha, R. G., (2007). Utilization of meat industry by-products: Protein hydrolysate from sheep visceral mass. *Bioresource Technology, 98*(2), 388–394.

Bhushan, S., Kalia, K., Sharma, M., Singh, B., & Ahuja, P. S., (2008). Processing of apple pomace for bioactive molecules. *Critical Reviews in Biotechnology, 28*(4), 285–296.

Bianco, A., Buiarelli, F., Cartoni, G., Coccioli, F., Jasionowska, R., & Margherita, P., (2003). Analysis by liquid chromatography-tandem mass spectrometry of biophenolic compounds in olives and vegetation waters, part I. *Journal of Separation Science, 26*(5), 409–416.

Bisignano, G., Tomaino, A., Cascio, R. L., Crisafi, G., Uccella, N., & Saija, A., (1999). On the in-vitro antimicrobial activity of oleuropein and hydroxytyrosol. *Journal of Pharmacy and Pharmacology, 51*(8), 971–974.

Botella, C., Diaz, A., De Ory, I., Webb, C., & Blandino, A., (2007). Xylanase and pectinase production by *Aspergillus* awamori on grape pomace in solid-state fermentation. *Process Biochemistry, 42*(1), 98–101.

Bouallagui, H., Touhami, Y., Cheikh, R. B., & Hamdi, M., (2005). Bioreactor performance in anaerobic digestion of fruit and vegetable wastes. *Process Biochemistry, 40*(3, 4), 989–995.

Breeze, P., & Breeze, P., (2018). Landfill waste disposal, anaerobic digestion, and energy production. *Energy from Waste,* 39–47.

Bugnicourt, E., Schmid, M., Nerney, O. M., & Wild, F., (2010). Wheylayer: The barrier coating of the future. *Coating International, 43*(11), 7–10.

Bulotta, S., Celano, M., Lepore, S. M., Montalcini, T., Pujia, A., & Russo, D., (2014). Beneficial effects of the olive oil phenolic components oleuropein and hydroxytyrosol: Focus on protection against cardiovascular and metabolic diseases. *Journal of Translational Medicine, 12*(1), 219.

Burange, A., Clark, J. H., & Luque, R., (2011). Trends in food and agricultural waste valorization. *Encyclopedia of Inorganic and Bioinorganic Chemistry,* 1–10.

Burton-Freeman, B. M., Sandhu, A. K., & Edirisinghe, I., (2017). Mangos and their bioactive components: Adding variety to the fruit plate for health. *Food & Function, 8*(9), 3010–3032.

Buzby, J. C., & Hyman, J., (2012). Total and per capita value of food loss in the United States. *Food Policy, 37*(5), 561–570.

Canakci, M., (2007). The potential of restaurant waste lipids as biodiesel feedstocks. *Bioresource Technology, 98*(1), 183–190.

Casa, R., D'Annibale, A., Pieruccetti, F., Stazi, S. R., Sermanni, G. G., & Cascio, B. L., (2003). Reduction of the phenolic components in olive-mill wastewater by an enzymatic treatment and its impact on durum wheat (*Triticum durum* Desf.) germinability. *Chemosphere, 50*(8), 959–966.

Cekmecelioglu, D., Demirci, A., Graves, R. E., & Davitt, N. H., (2005). Applicability of optimized in-vessel food waste composting for windrow systems. *Biosystems Engineering, 91*(4), 479–486.

Ćetković, G., Čanadanović-Brunet, J., Djilas, S., Savatović, S., Mandić, A., & Tumbas, V., (2008). Assessment of polyphenolic content and *in vitro* antiradical characteristics of apple pomace. *Food Chemistry, 109*(2), 340–347.

Chantaro, P., Devahastin, S., & Chiewchan, N., (2008). Production of antioxidant high dietary fiber powder from carrot peels. *LWT Food Sci. Technol., 41*(10), 1987–1994.

Char, B. L. N., & Azeemoddin, G., (1989). Edible fat from mango stones. *Acta Horticulturae.* Netherlands.

Chen, Y., Xiao, B., Chang, J., Fu, Y., Lv, P., & Wang, X., (2009). Synthesis of biodiesel from waste cooking oil using immobilized lipase in fixed bed reactor. *Energy Conversion and Management, 50*(3), 668–673.

Chhikara, N., Kushwaha, K., Sharma, P., Gat, Y., & Panghal, A., (2019). Bioactive compounds of beetroot and utilization in food processing industry: A critical review. *Food Chemistry, 272,* 192–200.

Cho, S. M., Gu, Y. S., & Kim, S. B., (2005). Extracting optimization and physical properties of yellowfin tuna (*Thunnus albacares*) skin gelatin compared to mammalian gelatins. *Food Hydrocolloids, 19*(2), 221–229.

Choi, Y. S., Kim, Y. B., Hwang, K. E., Song, D. H., Ham, Y. K., Kim, H. W., & Kim, C. J., (2016). Effect of apple pomace fiber and pork fat levels on quality characteristics of uncured, reduced-fat chicken sausages. *Poultry Science, 95*(6), 1465–1471.

Clementz, A., Torresi, P. A., Molli, J. S., Cardell, D., Mammarella, E., & Yori, J. C., (2019). Novel method for valorization of by-products from carrot discards. *LWT, 100*, 374–380.

Colaric, M., Veberic, R., Solar, A., Hudina, M., & Stampar, F., (2005). Phenolic acids, syringaldehyde, and juglone in fruits of different cultivars of *Juglans regia* L. *Journal of Agricultural and Food Chemistry, 53*(16), 6390–6396.

Coman, V., Teleky, B. E., Mitrea, L., Martău, G. A., Szabo, K., Călinoiu, L. F., & Vodnar, D. C., (2020). Bioactive potential of fruit and vegetable wastes. In: *Advances in Food and Nutrition Research* (Vol. 91, pp. 157–225). Academic Press.

Cordero, J. G., García-Escudero, R., Avila, J., Gargini, R., & García-Escudero, V., (2018). Benefit of oleuropein aglycone for Alzheimer's disease by promoting autophagy. *Oxidative Medicine and Cellular Longevity, 2018.*

Cossu, R. (2009). From triangles to cycles. *Waste Management (New York, NY), 29*(12), 2915–2917.

Couto, S. R., & Sanromán, M. A., (2006). Application of solid-state fermentation to food industry—A review. *Journal of Food Engineering, 76*(3), 291–302.

Curti, V., Di Lorenzo, A., Dacrema, M., Xiao, J., Nabavi, S. M., & Daglia, M., (2017). *In vitro* polyphenol effects on apoptosis: An update of literature data. In: *Seminars in Cancer Biology* (Vol. 46, pp. 119–131). Academic Press.

Das, S. P., Ravindran, R., Ahmed, S., Das, D., Goyal, D., Fontes, C. M., & Goyal, A., (2012). Bioethanol production involving recombinant C. thermocellum hydrolytic hemicellulase and fermentative microbes. *Applied Biochemistry and Biotechnology, 167*(6), 1475–1488.

De Almeida, V. F., García-Moreno, P. J., Guadix, A., & Guadix, E. M., (2015). Biodiesel production from mixtures of waste fish oil, palm oil and waste frying oil: Optimization of fuel properties. *Fuel Processing Technology, 133*, 152–160.

De Jesus, C. S. A., Ruth, V. G. E., Daniel, S. F. R., & Sharma, A., (2015). Biotechnological alternatives for the utilization of dairy industry waste products. *Advances in Bioscience and Biotechnology, 6*(03), 223.

Del, H. P., Rendueles, M., & Díaz, M., (2008). Effect of processing on functional properties of animal blood plasma. *Meat Science, 78*(4), 522–528.

Devatkal, S., Mendiratta, S. K., Kondaiah, N., Sharma, M. C., & Anjaneyulu, A. S. R., (2004). Physicochemical, functional, and microbiological quality of buffalo liver. *Meat Science, 68*(1), 79–86.

Dey, M. D., Das, S., Kumar, R., Doley, R., Bhattacharya, S. S., & Mukhopadhyay, R., (2017). Vermiremoval of methylene blue using *Eisenia fetida*: A potential strategy for bioremediation of synthetic dye-containing effluents. *Ecological Engineering, 106*, 200–208.

Dhillon, G. S., Kaur, S., & Brar, S. K., (2013). Perspective of apple processing wastes as low-cost substrates for bioproduction of high value products: A review. *Renewable and Sustainable Energy Reviews, 27*, 789–805.

Dimitrios, B., (2006). Sources of natural phenolic antioxidants. *Trends Food Sci. Technol., 17*, 505–512.

Doan, T. T., Henry-des-Tureaux, T., Rumpel, C., Janeau, J. L., & Jouquet, P., (2015). Impact of compost, vermicompost, and biochar on soil fertility, maize yield and soil erosion in Northern Vietnam: A 3 year mesocosm experiment. *Science of the Total Environment, 514*, 147–154.

Dominguez, J., & Edwards, C. A., (2010). Biology and ecology of earthworm species used for vermicomposting. *Vermiculture Technology: Earthworms, Organic Waste and Environmental Management*, 25–37.

Dorta, E., González, M., Lobo, M. G., Sánchez-Moreno, C., & De Ancos, B., (2014). Screening of phenolic compounds in by-product extracts from mangoes (*Mangifera indica* L.) by HPLC-ESI-QTOF-MS and multivariate analysis for use as a food ingredient. *Food Research International, 57*, 51–60.

Dorta, E., Lobo, M. G., & Gonzalez, M., (2012). Reutilization of mango byproducts: Study of the effect of extraction solvent and temperature on their antioxidant properties. *Journal of Food Science, 77*(1), C80–C88.

Dorta, P. E., Lobo, M. G., & González, M., (2013). Optimization of factors affecting extraction of antioxidants from mango seed. *Food and Bioprocess Technology, 6*, 1067–1081.

Du, G., Zhu, Y., Wang, X., Zhang, J., Tian, C., Liu, L., & Guo, Y., (2019). Phenolic composition of apple products and by-products based on cold pressing technology. *Journal of Food Science and Technology, 56*(3), 1389–1397.

Dubey, K. K., (2020). Food industry waste biorefineries: Future energy, valuable recovery, and waste treatment. In: *Refining Biomass Residues for Sustainable Energy and Bioproducts* (pp. 391–406). Academic Press.

Ejike, C. E., & Emmanuel, T. N., (2009). Cholesterol concentration in different parts of bovine meat sold in Nsukka, Nigeria: Implications for cardiovascular disease risk. *African Journal of Biochemistry Research, 3*(4), 95–97.

Encalada, A. M. I., Pérez, C. D., Flores, S. K., Rossetti, L., Fissore, E. N., & Rojas, A. M., (2019). Antioxidant pectin enriched fractions obtained from discarded carrots (*Daucus carota* L.) by ultrasound-enzyme assisted extraction. *Food Chemistry, 289*, 453–460.

Eriksson, M., Strid, I., & Hansson, P. A., (2014). Waste of organic and conventional meat and dairy products—A case study from Swedish retail. *Resources, Conservation, and Recycling, 83*, 44–52.

FAO, (2016). Food and Agriculture Organization of the United Nations. *Food Loss and Food Waste.* https://www.fao.org/publications/sofa/2016/en/ (accessed on 30 December 2021).

Feki, M., Allouche, N., Bouaziz, M., Gargoubi, A., & Sayadi, S., (2006). Effect of storage of olive mill wastewaters on hydroxytyrosol concentration. *European Journal of Lipid Science and Technology, 108*(12), 1021–1027.

Galanakis, C., (2020). Food waste valorization opportunities for different food industries. In: *The Interaction of Food Industry and Environment* (pp. 341–422). Academic Press.

Galhotra, K. K., & Wadhwa, B. K., (1993). Chemistry of ghee-residue, its significance and utilization. *Indian Journal of Dairy Science, 46*, 142–146.

Geerkens, C. H., Nagel, A., Just, K. M., Miller-Rostek, P., Kammerer, D. R., Schweiggert, R. M., & Carle, R., (2015). Mango pectin quality as influenced by cultivar, ripeness, peel particle size, blanching, drying, and irradiation. *Food Hydrocolloids, 51*, 241–251.

68   Integrated Waste Management Approaches for Food and Agricultural Byproducts

Germano, S., Pandey, A., Osaku, C. A., Rocha, S. N., & Soccol, C. R., (2003). Characterization and stability of proteases from *Penicillium* sp. produced by solid-state fermentation. *Enzyme and Microbial Technology, 32*(2), 246–251.

Ghosh, R., (2001). Fractionation of biological macromolecules using carrier phase ultrafiltration. *Biotechnology and Bioengineering, 74*(1), 1–11.

Girotto, F., Alibardi, L., & Cossu, R., (2015). Food waste generation and industrial uses: A review. *Waste Management, 45*, 32–41.

Gnanaprakasam, A., Sivakumar, V. M., Surendhar, A., Thirumarimurugan, M., & Kannadasan, T., (2013). Recent strategy of biodiesel production from waste cooking oil and process influencing parameters: A review. *Journal of Energy, 2013*.

Griffin, S., Sarfraz, M., Farida, V., Nasim, M. J., Ebokaiwe, A. P., Keck, C. M., & Jacob, C., (2018). No time to waste organic waste: Nanosizing converts remains of food processing into refined materials. *Journal of Environmental Management, 210*, 114–121.

Grosse, C., (1984). *Absatz und Vermarktungsmoglichkeiten fur Schlachtneben-produkte und Schlachtabfalle in der Bundesrepublik Deutschland* (Doctoral dissertation, Dissertation, Universitat Bonn, Institut fur Agrarpolitik, Marktforschung und Wirstschaftssoziologie, Marz).

Gülşen, A., Makris, D. P., & Kefalas, P., (2007). Biomimetic oxidation of quercetin: Isolation of a naturally occurring quercetin heterodimer and evaluation of its *in vitro* antioxidant properties. *Food Research International, 40*(1), 7–14.

Gustavsson, J., Cederberg, C., & Sonesson, U. (2011). Global Food Losses and Food Waste: Extent, Causes and Prevention. Study Conducted for the International Congress Save Food! At Interpack, Düsseldorf, Germany. *Food and Agriculture Organization of the United Nations,* Rome. http://www.fao.org/fileadmin/user_upload/suistainability/pdf/Global_Food_Losses_and_Food_Waste.pdf.

Hansen, C. L., & Cheong, D. Y., (2019). Agricultural waste management in food processing. In: *Handbook of Farm, Dairy, and Food Machinery Engineering* (pp. 673–716). Academic Press.

Hansen, V., Müller-Stöver, D., Munkholm, L. J., Peltre, C., Hauggaard-Nielsen, H., & Jensen, L. S., (2016). The effect of straw and wood gasification biochar on carbon sequestration, selected soil fertility indicators and functional groups in soil: An incubation study. *Geoderma, 269*, 99–107.

Harding, V. K., & Heale, J. B., (1980). Isolation and identification of the antifungal compounds accumulating in the induced resistance response of carrot root slices to *Botrytis cinerea. Physiological Plant Pathology, 17*(3), 277–289.

Harris, A. D., & Ramalingam, C., (2010). Xylanases and its application in food industry: A review. *Journal of Experimental Sciences*.

Hasan, M. M., Marzan, L. W., Hosna, A., Hakim, A., & Azad, A. K., (2017). Optimization of some fermentation conditions for the production of extracellular amylases by using *Chryseobacterium* and Bacillus isolates from organic kitchen wastes. *Journal of Genetic Engineering and Biotechnology, 15*(1), 59–68.

Hesas, R. H., Arami-Niya, A., Daud, W. M. A. W., & Sahu, J. N., (2013). Preparation and characterization of activated carbon from apple waste by microwave-assisted phosphoric acid activation: Application in methylene blue adsorption. *BioResources, 8*(2), 2950–2966.

Food Processing Byproducts

Ho, H., (2015). Xylanase production by Bacillus subtilis using carbon source of inexpensive agricultural wastes in two different approaches of submerged fermentation (SmF) and solid-state fermentation (SsF). *Journal of Food Processing and Technology, 6*(4).

Huc-Mathis, D., Journet, C., Fayolle, N., & Bose, V., (2019). Emulsifying properties of food by-products: Valorizing apple pomace and oat bran. *Colloids and Surfaces a-Physicochemical and Engineering Aspects, 568,* 84–91. https://doi.org/10.1016/j.colsurfa.2019.02.001.

Hufnagl, K., & Jensen-Jarolim, E., (2019). Does a carrot a day keep the allergy away?. *Immunology Letters, 206,* 54–58.

Irshad, A., Sureshkumar, S., Shalima, S. A., & Sutha, M., (2015). Slaughterhouse by-product utilization for sustainable meat industry: A review. *International Journal of Development Research, 5*(6), 4725–4734.

Jahirul, M. I., Rasul, M. G., Chowdhury, A. A., & Ashwath, N., (2012). Biofuels production through biomass pyrolysis—A technological review. *Energies, 5*(12), 4952–5001.

Jahurul, M. H. A., Zaidul, I. S. M., Ghafoor, K., Al-Juhaimi, F. Y., Nyam, K. L., Norulaini, N. A. N., & Omar, A. M., (2015). Mango (*Mangifera indica* L.) by-products and their valuable components: A review. *Food Chemistry, 183,* 173–180.

Jamilah, B., & Harvinder, K. G., (2002). Properties of gelatins from skins of fish—black tilapia (*Oreochromis mossambicus*) and red tilapia (*Oreochromis nilotica*). *Food Chemistry, 77*(1), 81–84.

Jayathilakan, K., Sultana, K., Radhakrishna, K., & Bawa, A. S., (2012). Utilization of byproducts and waste materials from meat, poultry, and fish processing industries: A review. *Journal of Food Science and Technology, 49*(3), 278–293.

Jitputti, J., Kitiyanan, B., Rangsunvigit, P., Bunyakiat, K., Attanatho, L., & Jenvanitpanjakul, P., (2006). Transesterification of crude palm kernel oil and crude coconut oil by different solid catalysts. *Chemical Engineering Journal, 116*(1), 61–66.

Julia, B. M., Belén, A. M., Georgina, B., & Beatriz, F., (2016). Potential use of soybean hulls and waste paper as supports in SSF for cellulase production by *Aspergillus Niger*. *Biocatalysis and Agricultural Biotechnology, 6,* 1–8.

Kähkönen, M. P., Heinämäki, J., Ollilainen, V., & Heinonen, M., (2003). Berry anthocyanins: Isolation, identification, and antioxidant activities. *Journal of the Science of Food and Agriculture, 83*(14), 1403–1411.

Kanazawa, K., & Sakakibara, H., (2000). High content of dopamine, a strong antioxidant, in cavendish banana. *Journal of Agricultural and Food Chemistry, 48*(3), 844–848.

Kapoor, M., Panwar, D., & Kaira, G. S., (2016). Bioprocesses for enzyme production using agro-industrial wastes: Technical challenges and commercialization potential. In: *Agro-Industrial Wastes as Feedstock for Enzyme Production* (pp. 61–93). Academic Press.

Karmee, S. K., & Lin, C. S. K., (2014). Valorization of food waste to biofuel: Current trends and technological challenges. *Sustainable Chemical Processes, 2*(1), 22.

Kattimani, V. R., Venkatesha, B. M., & Ananda, S., (2014). Biodiesel production from unrefined rice bran oil through three-stage transesterification. *Advances in Chemical Engineering and Science, 2014*.

Khan, M. A. A., Hossain, M. A., Hara, K., Osatomi, K., Ishihara, T., & Nozaki, Y., (2003). Effect of enzymatic fish-scrap protein hydrolysate on gel-forming ability and denaturation of lizardfish *Saurida wanieso* surimi during frozen storage. *Fisheries Science, 69*(6), 1271–1280.

Khedkar, R. D., & Singh, K., (2014). New Approaches for Food Industry Waste Utilization (pp. 51–65). *Biologix*, ISBN (81-88919).

Khiari, Z., Makris, D. P., & Kefalas, P., (2009). An investigation on the recovery of antioxidant phenolics from onion solid wastes employing water/ethanol-based solvent systems. *Food and Bioprocess Technology, 2*(4), 337.

Khosravi-Darani, K., Falahatpishe, H. R., & Jalali, M., (2008). Alkaline protease production on date waste by an alkalophilic *Bacillus* sp. 2–5 isolated from soil. *African Journal of Biotechnology, 7*(10).

Kibler, K. M., Reinhart, D., Hawkins, C., Motlagh, A. M., & Wright, J., (2018). Food waste and the food-energy-water nexus: A review of food waste management alternatives. *Waste Management, 74*, 52–62.

Kim, S. W., Park, D. K., & Kim, S. D., (2013). Pyrolytic characteristics of jatropha seed shell cake in thermobalance and fluidized bed reactors. *Korean Journal of Chemical Engineering, 30*(5), 1162–1170.

Kiran, E. U., Trzcinski, A. P., Ng, W. J., & Liu, Y., (2014). Enzyme production from food wastes using a biorefinery concept. *Waste and Biomass Valorization, 5*(6), 903–917.

Klass, D. L., (2004). Biomass for renewable energy and fuels. *Encyclopedia of Energy, 1*(1), 193–212.

Ko, E. Y., Nile, S. H., Sharma, K., Li, G. H., & Park, S. W., (2015). Effect of different exposed lights on quercetin and quercetin glucoside content in onion (*Allium cepa* L.). *Saudi Journal of Biological Sciences, 22*(4), 398–403.

Kojima, R., & Ishikawa, M., (2013). *Prevention and Recycling of Food Wastes in Japan: Policies and Achievements*. Kobe University, Japan, Resilient cities.

Koller, M., Bona, R., Chiellini, E., Fernandes, E. G., Horvat, P., Kutschera, C., & Braunegg, G., (2008). Polyhydroxyalkanoate production from whey by Pseudomonas hydrogenovora. *Bioresource Technology, 99*(11), 4854–4863.

Krishna, C., & Chandrasekaran, M., (1996). Banana waste as substrate for α-amylase production by *Bacillus subtilis* (CBTK 106) under solid-state fermentation. *Applied Microbiology and Biotechnology, 46*(2), 106–111.

Krishna, P. R., Srivastava, A. K., Ramaswamy, N. K., Suprasanna, P., & D'souza, S. F., (2012). Banana Peel as Substrate for α-Amylase Production Using Aspergillus Niger NCIM 616 and Process Optimization. *International Journal of Biotechnology, 11*, 314–319.

Kuddus, M., (2014). Bio-statistical approach for optimization of cold-active α-amylase production by novel psychrotolerant *M. foliorum* GA2 in solid-state fermentation. *Biocatalysis and Agricultural Biotechnology, 3*(2), 175–181.

Kulkarni, V. V., & Devatkal, S. K., (2015). Utilization of byproducts and waste materials from meat and poultry processing industry: A review. *Journal of Meat Science, 11*(1), 1–10.

Kumar, M. S., Rajiv, P., Rajeshwari, S., & Venckatesh, R., (2015). Spectroscopic analysis of vermicompost for determination of nutritional quality. *Spectrochimica Acta Part A: Molecular and Biomolecular Spectroscopy, 135*, 252–255.

Kumar, R. P., Gnansounou, E., Raman, J. K., & Baskar, G., (2019). *Refining Biomass Residues for Sustainable Energy and Bioproducts: Technology, Advances, Life Cycle Assessment, and Economics*. Academic Press.

Food Processing Byproducts 71

Labuckas, D. O., Maestri, D. M., Perello, M., Martínez, M. L., & Lamarque, A. L., (2008). Phenolics from walnut (*Juglans regia* L.) kernels: Antioxidant activity and interactions with proteins. *Food Chemistry, 107*(2), 607–612.

Lee, M., Lee, D., Cho, J., Kim, S., & Park, C., (2013). Enzymatic biodiesel synthesis in semi-pilot continuous process in near-critical carbon dioxide. *Applied Biochemistry and Biotechnology, 171*(5), 1118–1127.

Lee, S. U., Lee, J. H., Choi, S. H., Lee, J. S., Ohnisi-Kameyama, M., Kozukue, N., & Friedman, M., (2008). Flavonoid content in fresh, home-processed, and light-exposed onions and in dehydrated commercial onion products. *Journal of Agricultural and Food Chemistry, 56*(18), 8541–8548.

Lesage-Meessen, L., Navarro, D., Maunier, S., Sigoillot, J. C., Lorquin, J., Delattre, M., & Labat, M., (2001). Simple phenolic content in olive oil residues as a function of extraction systems. *Food Chemistry, 75*(4), 501–507.

Li, S., & Yang, X., (2016). Biofuel production from food wastes. In: *Handbook of Biofuels Production* (pp. 617–653). Woodhead Publishing.

Lin, C. S. K., Pfaltzgraff, L. A., Herrero-Davila, L., Mubofu, E. B., Abderrahim, S., Clark, J. H., & Luque, R., (2013). Food waste as a valuable resource for the production of chemicals, materials, and fuels. Current situation and global perspective. *Energy & Environmental Science, 6*(2), 426–464.

Liu, S., Jia, M., Chen, J., Wan, H., Dong, R., Nie, S., & Yu, Q., (2019). Removal of bound polyphenols and its effect on antioxidant and prebiotic properties of carrot dietary fiber. *Food Hydrocolloids, 93*, 284–292.

Loelillet, D., (1994). The European mango market: A promising tropical fruit. *Fruit, 49*, 332–334.

Lokman, I. M., Rashid, U., Yunus, R., & Taufiq-Yap, Y. H., (2014). Carbohydrate-derived solid acid catalysts for biodiesel production from low-cost feedstocks: A review. *Catalysis Reviews, 56*(2), 187–219.

Lowe, S. E., Theodorou, M. K., & Trinci, A. P., (1987). Cellulases and xylanase of an anaerobic rumen fungus grown on wheat straw, holocellulose, cellulose, and xylan. *Applied and Environmental Microbiology, 53*(6), 1216–1223.

Lu, Y., & Foo, L. Y., (1998). Constitution of some chemical components of apple seed. *Food Chemistry, 61*(1, 2), 29–33.

Luque, R., & Clark, J. H., (2013). Valorization of food residues: Waste to wealth using green chemical technologies. *Sustainable Chemical Processes, 1*(1), 1–3.

Ly, T. N., Hazama, C., Shimoyamada, M., Ando, H., Kato, K., & Yamauchi, R., (2005). Antioxidative compounds from the outer scales of onion. *Journal of Agricultural and Food Chemistry, 53*(21), 8183–8189.

Mahmood, T., & Hussain, S. T., (2010). Nanobiotechnology for the production of biofuels from spent tea. *African Journal of Biotechnology, 9*(6), 858–868.

Makris, D. P., (2015). Recovery and applications of enzymes from food wastes. In: *Food Waste Recovery* (pp. 361–379). Academic Press.

Manach, C., Mazur, A., & Scalbert, A., (2005). Polyphenols and prevention of cardiovascular diseases. *Current Opinions in Lipidology, 16*, 77–84.

Manyuchi, M. M., & Phiri, A., (2013). Vermicomposting in solid waste management: A review. *International Journal of Scientific Engineering and Technology, 2*(12), 1234–1242.

Manzano, S., & Williamson, G., (2010). Polyphenols and phenolic acids from strawberry and apple decrease glucose uptake and transport by human intestinal Caco-2 cells. *Molecular Nutrition & Food Research, 54*(12), 1773–1780.

Masoodi, F. A., Sharma, B., & Chauhan, G. S., (2002). Use of apple pomace as a source of dietary fiber in cakes. *Plant Foods for Human Nutrition, 57*(2), 121–128.

Mata-Alvarez, J., Dosta, J., Romero-Güiza, M. S., Fonoll, X., Peces, M., & Astals, S., (2014). A critical review on anaerobic co-digestion achievements between 2010 and 2013. *Renewable and Sustainable Energy Reviews, 36*, 412–427.

Maxwell, E. G., Belshaw, N. J., Waldron, K. W., & Morris, V. J., (2012). Pectin: An emerging new bioactive food polysaccharide. *Trends in Food Science & Technology, 24*(2), 64–73.

Melikoglu, M., Lin, C. S. K., & Webb, C., (2015). Solid-state fermentation of waste bread pieces by *Aspergillus awamori*: Analyzing the effects of airflow rate on enzyme production in packed bed bioreactors. *Food and Bioproducts Processing, 95*, 63–75.

Mesomo, M., Silva, M. F., Boni, G., Padilha, F. F., Mazutti, M., Mossi, A., & Treichel, H., (2009). Xanthan gum produced by *Xanthomonas campestris* from cheese whey: Production optimization and rheological characterization. *Journal of the Science of Food and Agriculture, 89*(14), 2440–2445.

Middleton, E., Kandaswami, C., & Theoharides, T. C., (2000). The effects of plant flavonoids on mammalian cells: Implications for inflammation, heart disease and cancer. *Pharmacological Reviews, 52*, 673–751.

Millati, R., Cahyono, R. B., Ariyanto, T., Azzahrani, I. N., Putri, R. U., & Taherzadeh, M. J., (2019). Agricultural, industrial, municipal, and forest wastes: An overview. In: *Sustainable Resource Recovery and Zero Waste Approaches* (pp. 1–22). Elsevier.

Moral, A., Reyero, I., Alfaro, C., Bimbela, F., & Gandía, L. M., (2018). Syngas production by means of biogas catalytic partial oxidation and dry reforming using Rh-based catalysts. *Catalysis Today, 299*, 280–288.

Moure, A., Cruz, J. M., Franco, D., Domínguez, J. M., Sineiro, J., Domínguez, H., Núñez, M. J., & Parajó, J. C., (2001). Natural antioxidants from residual sources: A review. *Food Chem., 72*, 145–171.

Mourtzinos, I., Prodromidis, P., Grigorakis, S., Makris, D. P., Biliaderis, C. G., & Moschakis, T., (2018). Natural food colorants derived from onion wastes: Application in a yoghurt product. *Electrophoresis, 39*(15), 1975–1983.

Mullen, A. M., Alvarez, C., Pojic, M., Hadnadev, T. D., & Papageorgiou, M., (2015). Chapter 2. Classification and target compounds. In: Galanakis, C. M., (ed.), *Food Waste Recovery: Processing Technologies and Industrial Techniques.*

Murthy, P. S., Madhava, N. M., & Srinivas, P., (2009). Production of α-amylase under solid-state fermentation utilizing coffee waste. *Journal of Chemical Technology & Biotechnology: International Research in Process, Environmental & Clean Technology, 84*(8), 1246–1249.

Naik, S. N., Goud, V. V., Rout, P. K., & Dalai, A. K., (2010). Production of first and second-generation biofuels: A comprehensive review. *Renewable and Sustainable Energy Reviews, 14*(2), 578–597.

Nair, S. V., & Rupasinghe, H. V., (2014). Fatty acid esters of phloridzin induce apoptosis of human liver cancer cells through altered gene expression. *PLoS One, 9*(9), e107149.

Narra, M., Dixit, G., Divecha, J., Madamwar, D., & Shah, A. R., (2012). Production of cellulases by solid-state fermentation with *Aspergillus terreus* and enzymatic hydrolysis of mild alkali-treated rice straw. *Bioresource Technology, 121*, 355–361.

Nemerow, N. L., & Agardy, F. J., (1998). *Strategies of Industrial and Hazardous Waste Management*. John Wiley & Sons.

Nile, A., Nile, S. H., Kim, D. H., Keum, Y. S., Seok, P. G., & Sharma, K., (2018). Valorization of onion solid waste and their flavonols for assessment of cytotoxicity, enzyme inhibitory and antioxidant activities. *Food and Chemical Toxicology, 119*, 281–289.

Nile, S. H., Nile, A. S., Keum, Y. S., & Sharma, K., (2017). Utilization of quercetin and quercetin glycosides from onion (*Allium cepa* L.) solid waste as an antioxidant, urease, and xanthine oxidase inhibitors. *Food Chemistry, 235*, 119–126.

Obi, F. O., Ugwuishiwu, B. O., & Nwakaire, J. N., (2016). Agricultural waste concept, generation, utilization, and management. *Nigerian Journal of Technology, 35*(4), 957–964.

Oliveira, F., Souza, C. E., Peclat, V. R., Salgado, J. M., Ribeiro, B. D., Coelho, M. A., & Belo, I., (2017). Optimization of lipase production by *Aspergillus ibericus* from oil cakes and its application in esterification reactions. *Food and Bioproducts Processing, 102*, 268–277.

Omar, S. H., (2010). Oleuropein in olive and its pharmacological effects. *Scientia Pharmaceutica, 78*(2), 133–154.

Omar, W. N. N. W., & Amin, N. A. S., (2011). Optimization of heterogeneous biodiesel production from waste cooking palm oil via response surface methodology. *Biomass and Bioenergy, 35*(3), 1329–1338.

Onwulata, C., & Huth, P., (2009). *Whey Processing, Functionality, and Health Benefits* (Vol. 82). John Wiley & Sons.

Panda, S. K., Mishra, S. S., Kayitesi, E., & Ray, R. C., (2016). Microbial-processing of fruit and vegetable wastes for production of vital enzymes and organic acids: Biotechnology and scopes. *Environmental Research, 146*, 161–172.

Pandey, A., Nigam, P., Soccol, C. R., Soccol, V. T., Singh, D., & Mohan, R., (2000). Advances in microbial amylases. *Biotechnology and Applied Biochemistry, 31*(2), 135–152.

Pandian, S. R., Deepak, V., Kalishwaralal, K., Rameshkumar, N., Jeyaraj, M., & Gurunathan, S., (2010). Optimization and fed-batch production of PHB utilizing dairy waste and seawater as nutrient sources by Bacillus megaterium SRKP-3. *Bioresource Technology, 101*(2), 705–711.

Panesar, P. S., Kumari, S., & Panesar, R., (2013). Biotechnological approaches for the production of prebiotics and their potential applications. *Critical Reviews in Biotechnology, 33*(4), 345–364.

Panzella, L., Moccia, F., Nasti, R., Marzorati, S., Verotta, L., & Napolitano, A., (2020). Bioactive phenolic compounds from agri-food wastes: An update on green and sustainable extraction methodologies. *Frontiers in Nutrition, 7*.

Papanikolaou, S., Dimou, A., Fakas, S., Diamantopoulou, P., Philippoussis, A., Galiotou-Panayotou, M., & Aggelis, G., (2011). Biotechnological conversion of waste cooking olive oil into lipid-rich biomass using Aspergillus and Penicillium strains. *Journal of Applied Microbiology, 110*(5), 1138–1150.

Parfitt, J., Barthel, M., & Macnaughton, S., (2010). Food waste within food supply chains: Quantification and potential for change to 2050. *Philosophical Transactions of the Royal Society B: Biological Sciences, 365*(1554), 3065–3081.

Pattanaik, L., Pattnaik, F., Saxena, D. K., & Naik, S. N., (2019). Biofuels from agricultural wastes. In: *Second and Third Generation of Feedstocks* (pp. 103–142). Elsevier.

Pazmiño-Durán, E. A., Giusti, M. M., Wrolstad, R. E., & Glória, M. B. A., (2001). Anthocyanins from banana bracts (Musa X paradisiaca) as potential food colorants. *Food Chemistry, 73*(3), 327–332.

Peng, H., Deng, Z., Chen, X., Sun, Y., Zhang, B., & Li, H., (2018). Major chemical constituents and antioxidant activities of different extracts from the peduncles of *Hovenia acerba* Lindl. *Int. J. Food Prop., 21*(1), 2135–2155.

Pereira, M. D. G., Cardoso De, S. N. L., Fontes, M. P. F., Souza, A. N., Carvalho, M. T., De Lima, S. R., & Verónica, F. N. R. A., (2014). An overview of the environmental applicability of vermicompost: From wastewater treatment to the development of sensitive analytical methods. *The Scientific World Journal, 2014.*

Pérez-Gregorio, M. R., Regueiro, J., Simal-Gándara, J., Rodrigues, A. S., & Almeida, D. P. F., (2014). Increasing the added-value of onions as a source of antioxidant flavonoids: A critical review. *Critical Reviews in Food Science and Nutrition, 54*(8), 1050–1062.

Peschel, W., Sánchez-Rabaneda, F., Diekmann, W., Plescher, A., Gartzía, I., Jiménez, D., & Codina, C., (2006). An industrial approach in the search of natural antioxidants from vegetable and fruit wastes. *Food Chemistry, 97*(1), 137–150.

Petkovšek, M. M., Stampar, F., & Veberic, R., (2008). Increased phenolic content in apple leaves infected with the apple scab pathogen. *Journal of Plant Pathology*, 49–55.

Pierson, J. T., Monteith, G. R., Roberts-Thomson, S. J., Dietzgen, R. G., Gidley, M. J., & Shaw, P. N., (2014). Phytochemical extraction, characterization, and comparative distribution across four mango (*Mangifera indica* L.) fruit varieties. *Food Chemistry, 149*, 253–263.

Pinto, S., Bhatt, J. D., & Prajapati, J. P., (2014). Evaluation of selected emulsifiers and buttermilk in the manufacture of reduced-fat Paneer. *Basic Research Journal of Food Science and Technology, 1*(4), 1–14.

Pleissner, D., Lam, W. C., Sun, Z., & Lin, C. S. K., (2013). Food waste as a nutrient source in heterotrophic microalgae cultivation. *Bioresource Technology, 137*, 139–146.

Prakash, B., Vidyasagar, M., Madhukumar, M. S., Muralikrishna, G., & Sreeramulu, K., (2009). Production, purification, and characterization of two extremely halotolerant, thermostable, and alkali-stable α-amylases from chromohalobacter sp. TVSP 101. *Process Biochemistry, 44*(2), 210–215.

Pramanik, P., & Chung, Y. R., (2011). Changes in fungal population of fly ash and vinasse mixture during vermicomposting by *Eudrilus eugeniae* and *Eisenia fetida*: Documentation of cellulase isozymes in vermicompost. *Waste Management, 31*(6), 1169–1175.

Puravankara, D., Boghra, V., & Sharma, R. S., (2000). Effect of antioxidant principles isolated from mango (*Mangifera indica* L) seed kernels on oxidative stability of buffalo ghee (butter-fat). *Journal of the Science of Food and Agriculture, 80*(4), 522–526.

Püssa, T., Floren, J., Kuldkepp, P., & Raal, A., (2006). Survey of grapevine Vitis vinifera stem polyphenols by liquid chromatography– diode array detection– tandem mass spectrometry. *Journal of Agricultural and Food Chemistry, 54*(20), 7488–7494.

Puupponen-Pimia, R., Nohynek, L., Meier, C., Ka"hko"nen, M., Heinonen, M., Hopia, A., et al., (2001). Antimicrobial properties of phenolic compounds from berries. *Journal of Applied Microbiology, 90*, 494–507.

Quaik, S., & Ibrahim, M. H., (2013). A review on potential of vermicomposting derived liquids in agricultural use. *International Journal of Scientific and Research Publications, 3*(3), 1–6.

Rafiq, S. M., & Rafiq, S. I., (2019). Milk by-products utilization. In: *Current Issues and Challenges in the Dairy Industry*. IntechOpen.

Ramos-Cormenzana, A., Juarez-Jimenez, B., & Garcia-Pareja, M. P., (1996). Antimicrobial activity of olive mill wastewaters (alpechin) and biotransformed olive oil mill wastewater. *International Biodeterioration & Biodegradation, 38*(3, 4), 283–290.

Ramrakhiani, L., & Chand, S., (2011). Recent progress on phospholipases: Different sources, assay methods, industrial potential and pathogenicity. *Applied Biochemistry and Biotechnology, 164*(7), 991–1022.

Rather, S. A., Akhter, R., Masoodi, F. A., Gani, A., & Wani, S. M., (2015). Utilization of apple pomace powder as a fat replacer in goshtaba: A traditional meat product of Jammu and Kashmir, India. *Journal of Food Measurement and Characterization, 9*(3), 389–399.

Ravindran, R., & Jaiswal, A. K., (2016). Exploitation of food industry waste for high-value products. *Trends in Biotechnology, 34*(1), 58–69.

Ravindran, R., Hassan, S. S., Williams, G. A., & Jaiswal, A. K., (2018). A review on bioconversion of agro-industrial wastes to industrially important enzymes. *Bioengineering, 5*(4), 93.

Ren, Y., Yu, M., Wu, C., Wang, Q., Gao, M., Huang, Q., & Liu, Y., (2018). A comprehensive review on food waste anaerobic digestion: Research updates and tendencies. *Bioresource Technology, 247*, 1069–1076.

Robles-Sánchez, R. M., Rojas-Graü, M. A., Odriozola-Serrano, I., González-Aguilar, G., & Martin-Belloso, O., (2013). Influence of alginate-based edible coating as carrier of anti-browning agents on bioactive compounds and antioxidant activity in fresh-cut Kent mangoes. *LWT-Food Science and Technology, 50*(1), 240–246.

Rorat, A., & Vandenbulcke, F., (2019). Earthworms converting domestic and food industry wastes into biofertilizer. In: *Industrial and Municipal Sludge* (pp. 83–106). Butterworth-Heinemann.

Rosales-Mendoza, S., & Tello-Olea, M. A., (2015). Carrot cells: A pioneering platform for biopharmaceuticals production. *Molecular Biotechnology, 57*(3), 219–232.

Sadh, P. K., Duhan, S., & Duhan, J. S., (2018). Agro-industrial wastes and their utilization using solid-state fermentation: A review. *Bioresources and Bioprocessing, 5*(1), 1.

Sahnoun, M., Kriaa, M., Elgharbi, F., Ayadi, D. Z., Bejar, S., & Kammoun, R., (2015). *Aspergillus Niger* S2 alpha-amylase production under solid-state fermentation: Optimization of culture conditions. *International Journal of Biological Macromolecules, 75*, 73–80.

Sakuragi, K., Li, P., Otaka, M., & Makino, H., (2016). Recovery of bio-oil from industrial food waste by liquefied dimethyl ether for biodiesel production. *Energies, 9*(2), 106.

Salemdeeb, R., Zu Ermgassen, E. K., Kim, M. H., Balmford, A., & Al-Tabbaa, A., (2017). Environmental and health impacts of using food waste as animal feed: A comparative analysis of food waste management options. *Journal of Cleaner Production, 140*, 871–880.

Salihu, A., Bala, M., & Alam, M. Z., (2016). Lipase production by *Aspergillus Niger* using shea nut cake: An optimization study. *Journal of Taibah University for Science, 10*(6), 850–859.

Samman, S., Lyons, W. P. M., & Cook, N. C., (1998). Flavonoids and coronary heart disease: Dietary perspectives. In: Rice-Evans, C. A., & Packer, L., (eds.), *Flavonoids in Health and Disease* (pp. 469–482). New York: Marcel Dekker.

Sánchez-Arreola, E., Martin-Torres, G., Lozada-Ramírez, J. D., Hernández, L. R., Bandala-González, E. R., & Bach, H., (2015). Biodiesel production and de-oiled seed cake nutritional values of a Mexican edible *Jatropha curcas*. *Renewable Energy, 76*, 143–147.

Sánchez-Rabaneda, F., Jauregui, O., Lamuela-Raventós, R. M., Viladomat, F., Bastida, J., & Codina, C., (2004). Qualitative analysis of phenolic compounds in apple pomace using liquid chromatography coupled to mass spectrometry in tandem mode. *Rapid Communications in Mass Spectrometry, 18*(5), 553–563.

Sawant, R., & Nagendran, S., (2014). Protease: An enzyme with multiple industrial applications. *World Journal of Pharmacy and Pharmaceutical Sciences, 3*(6), 568–579.

Sayed, M. E., Askar, A. A., Hamzawi, L. F., Fatma, F. A., Mohamed, A. G., & El Sayed, S. M., (2010). Utilization of buttermilk concentrate in the manufacture of functional processed cheese spread. *Journal of American Science, 6*(9), 876–882.

Scano, E. A., Asquer, C., Pistis, A., Ortu, L., Demontis, V., & Cocco, D., (2014). Biogas from anaerobic digestion of fruit and vegetable wastes: Experimental results on pilot-scale and preliminary performance evaluation of a full-scale power plant. *Energy Conversion and Management, 77*, 22–30.

Schieber, A., (2017). Side streams of plant food processing as a source of valuable compounds: Selected examples. *Annual Review of Food Science and Technology, 8*, 97–112.

Schieber, A., Berardini, N., & Carle, R., (2003). Identification of flavonol and xanthone glycosides from mango (*Mangifera indica* L. Cv. "Tommy Atkins") peels by high-performance liquid chromatography-electrospray ionization mass spectrometry. *Journal of Agricultural and Food Chemistry, 51*(17), 5006–5011.

Schieber, A., Keller, P., & Carle, R., (2001). Determination of phenolic acids and flavonoids of apple and pear by high-performance liquid chromatography. *Journal of Chromatography A, 910*(2), 265–273.

Schieber, A., Stintzing, F. C., & Carle, R., (2001b). By-products of plant food processing as a source of functional compounds—Recent developments. *Trends Food Sci. Technol., 12*, 401–413.

Schulze-Kaysers, N., Feuereisen, M. M., & Schieber, A., (2015). Phenolic compounds in edible species of the *Anacardiaceae* family: A review. *RSC Advances, 5*(89), 73301–73314.

Sen, S. K., Dora, T. K., Bandyopadhyay, B., Mohapatra, P. K. D., & Raut, S., (2014). Thermostable alpha-amylase enzyme production from hot spring isolates *Alcaligenes faecalis* SSB17–statistical optimization. *Biocatalysis and Agricultural Biotechnology, 3*(4), 218–226.

Serna-Cock, L., Torres-León, C., & Ayala-Aponte, A., (2015). Evaluation of food powders obtained from peels of mango (Mangifera indica) as sources of functional ingredients. *Inform Technol., 26*, 41–50.

Seyis, I., & Aksoz, N., (2005). Xylanase Production from Trichoderma Harzianum 1073 D3 with Alternative Carbon and Nitrogen Sources. *Food Technology & Biotechnology, 43*(1), 37–40.

Shahidi, F., & Peng, H., (2018). Bioaccessibility and bioavailability of phenolic compounds. *J. Food Bioact., 4*, 11–68.

Shahidi, F., Varatharajan, V., Oh, W. Y., & Peng, H., (2019). Phenolic compounds in agrifood by-products, their bioavailability and health effects. *J. Food Bioact., 5*(1), 57–119.

Shahidi, F., Zhong, Y., Wijeratne, S. S. K., & Ho, C. T., (2009). Almon and almond products: Nutraceutical components and health effects. In: Alasalvar, C., & Shahidi, F., (eds.), *Tree Nuts: Composition, Phytochemicals, and Health Effects* (pp. 127–138). CRC press, Boca Raton, FL.

Sharma, B. D., (2011). *Outlines of Meat Science and Technology, 2011.* Jaypee Brothers Medical.

Sharma, K. D., Karki, S., Thakur, N. S., & Attri, S., (2012). Chemical composition, functional properties and processing of carrot—A review. *Journal of Food Science and Technology, 49*(1), 22–32.

Sharma, R., Oberoi, H. S., & Dhillon, G. S., (2016). Fruit and vegetable processing waste: Renewable feedstocks for enzyme production. In: *Agro-Industrial Wastes as Feedstock for Enzyme Production* (pp. 23–59). Academic Press.

Shih, J. C., (1993). Recent development in poultry waste digestion and feather utilization—A review. *Poultry Science, 72*(9), 1617–1620.

Shore, M., Broughton, N. W., & Bunstead, N., (1984). Anaerobic treatment of wastewaters in the beet sugar industry. *Water Pollut. Control* (Vol. 83, No. 4). Maidstone; United Kingdom.

Sila, D. N., Van, B. S., Duvetter, T., Fraeye, I., De Roeck, A., Van, L. A., & Hendrickx, M., (2009). Pectins in processed fruits and vegetables: Part II—structure-function relationships. *Comprehensive Reviews in Food Science and Food Safety, 8*(2), 86–104.

Silva, R. V., De Brito, J., Lynn, C. J., & Dhir, R. K., (2019). Environmental impacts of the use of bottom ashes from municipal solid waste incineration: A review. *Resources, Conservation, and Recycling, 140*, 23–35.

Sindhu, R., Gnansounou, E., Rebello, S., Binod, P., Varjani, S., Thakur, I. S., & Pandey, A., (2019). Conversion of food and kitchen waste to value-added products. *Journal of Environmental Management, 241*, 619–630.

Singh, B. N., Singh, B. R., Singh, R. L., Prakash, D., Singh, D. P., Sarma, B. K., & Singh, H. B., (2009). Polyphenolics from various extracts/fractions of red onion (*Allium cepa*) peel with potent antioxidant and antimutagenic activities. *Food and Chemical Toxicology, 47*(6), 1161–1167.

Sirisakulwat, S., Sruamsiri, P., Carle, R., & Neidhart, S., (2010). Resistance of industrial mango peel waste to pectin degradation prior to by-product drying. *International Journal of Food Science & Technology, 45*(8), 1647–1658.

Soler-Rivas, C., Espín, J. C., & Wichers, H. J., (2000). Oleuropein and related compounds. *Journal of the Science of Food and Agriculture, 80*(7), 1013–1023.

Soliman, H. M., Sherief, A. A., & EL-Tanash, A. B., (2012). Production of xylanase by *Aspergillus Niger* and *Trichoderma viride* using some agriculture residues. *International Journal of Agricultural Research, 7*(1), 46–57.

Solís-Fuentes, J. A., & Durán-de-Bazúa, M. D. C., (2011). Mango (*Mangifera indica* L.) seed and its fats. In: *Nuts and Seeds in Health and Disease Prevention* (pp. 741–748). Academic Press.

Soni, R., (2013). *Production, Purification, and Industrial Applications of Cellulase from Aspergillus sp.* AMA.

Soong, Y. Y., Barlow, P. J., & Perera, C. O., (2004). A cocktail of phytonutrients: Identification of polyphenols, phytosterol, and tocopherols from mango seed kernel. In: *IFT Annual Meeting* (pp. 12–16).

Souquet, J. M., Cheynier, V., Brossaud, F., & Moutounet, M., (1996). Polymeric proanthocyanidins from grape skins. *Phytochemistry, 43*(2), 509–512.

Suthar, S., (2009). Vermicomposting of vegetable-market solid waste using *Eisenia fetida*: Impact of bulking material on earthworm growth and decomposition rate. *Ecological Engineering, 35*(5), 914–920.

Tan, Y. H., Abdullah, M. O., & Nolasco-Hipolito, C., (2015). The potential of waste cooking oil-based biodiesel using heterogeneous catalyst derived from various calcined eggshells coupled with an emulsification technique: A review on the emission reduction and engine performance. *Renewable and Sustainable Energy Reviews, 47*, 589–603.

Thomas, L., Larroche, C., & Pandey, A., (2013). Current developments in solid-state fermentation. *Biochemical Engineering Journal, 81*, 146–161.

Tomás-Barberán, F. A., Llorach, R., Espín, J. C., & Ferreres, F., (2004). Agrifood residues as a source of phytochemicals. In: Keith, W., Craig, F., & Andrew, S., (eds.), *Total Food* (pp. 42–48). Institute of Food Research, Proceedings.

Toor, R. K., & Savage, G. P., (2005). Antioxidant activity in different fractions of tomatoes. *Food Research International, 38*(5), 487–494.

Torres-León, C., Ramírez-Guzman, N., Londoño-Hernandez, L., Martinez-Medina, G. A., Díaz-Herrera, R., Navarro-Macias, V., & Aguilar, C. N., (2018). Food waste and byproducts: An opportunity to minimize malnutrition and hunger in developing countries. *Frontiers in Sustainable Food Systems, 2*, 52.

Torres-León, C., Rojas, R., Contreras-Esquivel, J. C., Serna-Cock, L., Belmares-Cerda, R. E., & Aguilar, C. N., (2016). Mango seed: Functional and nutritional properties. *Trends in Food Science & Technology, 55*, 109–117.

Torres-León, C., Rojas, R., Serna-Cock, L., Belmares-Cerda, R., & Aguilar, C. N., (2017). Extraction of antioxidants from mango seed kernel: Optimization assisted by microwave. *Food and Bioproducts Processing, 105*, 188–196.

Torres-León, C., Vicente, A. A., Flores-López, M. L., Rojas, R., Serna-Cock, L., Alvarez-Pérez, O. B., & Aguilar, C. N., (2018). Edible films and coatings based on mango (var. Ataulfo) by-products to improve gas transfer rate of peach. *LWT, 97*, 624–631.

Trninić, M., Jovović, A., & Stojiljković, D., (2016). A steady-state model of agricultural waste pyrolysis: A mini-review. *Waste Management & Research, 34*(9), 851–865.

Turksoy, S., & Özkaya, B., (2011). Pumpkin and carrot pomace powders as a source of dietary fiber and their effects on the mixing properties of wheat flour dough and cookie quality. *Food Science and Technology Research, 17*(6), 545–553.

Tyug, T. S., Johar, M. H., & Ismail, A., (2010). Antioxidant properties of fresh, powder, and fiber products of mango (*Mangifera foetida*) fruit. *International Journal of Food Properties, 13*(4), 682–691.

Food Processing Byproducts

79

United Nations University, & World Health Organization, (2004). *Human Energy Requirements: Report of a Joint FAO/WHO/UNU Expert Consultation: Rome* (Vol. 1). Food & Agriculture Org.

Ünsal, M., & Aktaş, N., (2003). Fractionation and characterization of edible sheep tail fat. *Meat Science, 63*(2), 235–239.

Veerabhadrappa, M. B., Shivakumar, S. B., & Devappa, S., (2014). Solid-state fermentation of jatropha seed cake for optimization of lipase, protease, and detoxification of anti-nutrients in jatropha seed cake using *Aspergillus* versicolor CJS-98. *Journal of Bioscience and Bioengineering, 117*(2), 208–214.

Velioglu, Y. S., Mazza, G., Gao, L., & Oomah, B. D., (1998). Antioxidant activity and total phenolics in selected fruits, vegetables, and grain products. *Journal of Agricultural and Food Chemistry, 46*(10), 4113–4117.

Veloorvalappil, N. J., Robinson, B. S., Selvanesan, P., Sasidharan, S., Kizhakkepawothail, N. U., Sreedharan, S., & Sailas, B., (2013). Versatility of microbial proteases. *Advances in Enzyme Research, 2013.*

Vendruscolo, F., Albuquerque, P. M., Streit, F., Esposito, E., & Ninow, J. L., (2008). Apple pomace: A versatile substrate for biotechnological applications. *Critical Reviews in Biotechnology, 28*(1), 1–12.

Verma, B. B., & De, S., (1978). Preparation of chocsidu-burfi from ghee-residue. *Indian Journal of Dairy Science.*

Viniegra-González, G., Favela-Torres, E., Aguilar, C. N., De Rómero-Gomez, S. J., Diaz-Godinez, G., & Augur, C., (2003). Advantages of fungal enzyme production in solid-state over liquid fermentation systems. *Biochemical Engineering Journal, 13*(2, 3), 157–167.

Visioli, F., Bellosta, S., & Galli, C., (1998). Oleuropein, the bitter principle of olives, enhances nitric oxide production by mouse macrophages. *Life Sciences, 62*(6), 541–546.

Wadhwa, M., & Bakshi, M. P. S., (2013). *Utilization of Fruit and Vegetable Wastes as Livestock Feed and as Substrates for Generation of other Value-Added Products, 4,* 1–67. Rap Publication.

Wan, Y., Ghosh, R., & Cui, Z., (2002). High-resolution plasma protein fractionation using ultrafiltration. *Desalination, 144*(1–3), 301–306.

Wang, Q., Wang, X., Wang, X., & Ma, H., (2008). Glucoamylase production from food waste by *Aspergillus Niger* under submerged fermentation. *Process Biochemistry, 43*(3), 280–286.

Wang, X., Kristo, E., & LaPointe, G., (2019). The effect of apple pomace on the texture, rheology, and microstructure of set type yogurt. *Food Hydrocolloids, 91,* 83–91.

Wijngaard, H. H., Christian, R., & Nigel, B., (2009). A survey of Irish fruit and vegetable waste and by-products as a source of polyphenolic antioxidants. *Food Chem., 116,* 202–207.

Wolfe, K. L., & Liu, R. H., (2003). Apple peels as a value-added food ingredient. *Journal of Agricultural and Food Chemistry, 51*(6), 1676–1683.

Wolfe, K., Wu, X., & Liu, R. H., (2003). Antioxidant activity of apple peels. *Journal of Agricultural and Food Chemistry, 51*(3), 609–614.

Yaakob, Z., Mohammad, M., Alherbawi, M., Alam, Z., & Sopian, K., (2013). Overview of the production of biodiesel from waste cooking oil. *Renewable and Sustainable Energy Reviews, 18,* 184–193.

Yadav, S., Malik, A., Pathera, A., Islam, R. U., & Sharma, D., (2016). Development of dietary fibre enriched chicken sausages by incorporating corn bran, dried apple pomace and dried tomato pomace. *Nutrition & Food Science.*

Yaman, S., (2004). Pyrolysis of biomass to produce fuels and chemical feedstocks. *Energy Conversion and Management, 45,* 651–671.

Yi, C., Shi, J., Kramer, J., Xue, S., Jiang, Y., Zhang, M., & Pohorly, J., (2009). Fatty acid composition and phenolic antioxidants of winemaking pomace powder. *Food Chemistry, 114*(2), 570–576.

Zhang, D., & Hamauzu, Y., (2004). Phenolic compounds and their antioxidant properties in different tissues of carrots (*Daucus carota* L.). *Journal of Food Agriculture and Environment, 2,* 95–100.

Zhang, Y., Wong, W. T., & Yung, K. F., (2013). One-step production of biodiesel from rice bran oil catalyzed by chlorosulfonic acid modified zirconia via simultaneous esterification and transesterification. *Bioresource Technology, 147,* 59–64.

Zhou, J., Kang, L., Wang, H. W., Yang, T., & Dai, C. C., (2014). Liquid laccase production by *Phomopsis* liquidambari B3 accelerated phenolic acids degradation in long-term cropping soil of peanut. *Acta Agriculturae Scandinavica, Section B—Soil & Plant Science, 64*(8), 683–693.

CHAPTER 3

Waste to Wealth: Reduction, Reuse, and Recycling of Food and Agricultural Waste

RUBIYA RASHID,[1] F. A. MASOODI,[1] SAJAD MOHD WANI,[2] SHAKEEL AHMAD BHAT,[3] SHAZIYA MANZOOR,[1] OMAR BASHIR,[2] ROUF AHMAD BHAT,[4] and AB. WAHEED WANI[5]

[1] *Department of Food Science and Technology, University of Kashmir, Srinagar, Jammu and Kashmir, India, E-mail: rubiyarashideng@gmail.com (Rubiya Rashid)*

[2] *Department of Food Technology and Nutrition, Lovely Professional University, Punjab, India*

[3] *College of Agricultural Engineering, Sher-e-Kashmir University of Agricultural Sciences and Technology, Shalimar, Srinagar, Jammu and Kashmir, India*

[4] *Department of Environmental Science, Sheri-e-Kashmir University of Agricultural Sciences and Technology, Shalimar, Srinagar, Jammu and Kashmir, India*

[5] *Department of Fruit Science, Sheri-e-Kashmir University of Agricultural Sciences and Technology, Shalimar, Srinagar, India*

Integrated Waste Management Approaches for Food and Agricultural Byproducts. Tawheed Amin, PhD, Omar Bashir, Shakeel Ahmad Bhat & Muneeb Ahmad Malik (Eds.)
© 2023 Apple Academic Press, Inc. Co-published with CRC Press (Taylor & Francis)

ABSTRACT

Waste to wealth concept practiced in many developed countries envisages waste as a resource generator and focus on identification, development, and deployment of technologies to treat waste to generate energy, recycle material and extract resource of value from it. This notion will work to support the evolution of innovative technologies that will originate clean and green environment. With escalation of concern about global food security and environmental impacts, in particular resource depletion and greenhouse gas emission allocated to agricultural waste have build up attention to subject matter. 3R strategy helps in reduction, reuse, and recycling of food and agricultural waste globally, which creates well balance for nutritional security and environmental sustainability. In this chapter, various strategies for conversion of agricultural waste to wealth have been discussed to identify and use it as a new business opportunity and economy generation.

3.1 INTRODUCTION

The Food and Agricultural Organization of the United Nations (FAO) reported that about 1.3 billion tons of food is lost or wasted every year, which is around one-third of all global food production. These food wastes are produced from households, agricultural residues, commercial, and industrial sectors. Food waste and loss are frequently used interchangeably to identify substances which were considered for human consumption, but eventually get degraded, contaminated, lost or discharged. Food and Agricultural Organization, however, described FL as any change in the availability, edibility, wholesomeness or quality of consumable material that prevents people from consuming it, while food waste can be described as the FL that occurs mostly at the retail and final consumption stages and its development is linked to the actions of retailers and consumers (Girotto et al., 2015). Food processing industry produces FL and waste throughout the supply chain such as damage during transportation, during storage, processing, packaging, in retail systems and also in markets. Food waste and losses represent besides being wasting of that food commodity but also the wastage of those resources which were used for its production such as land for cultivation, water for irrigation, fertilizers, chemicals, energy for transportation and labor. Natural resource degradation used during farming operations not only exists, but had an impact on social, economic, and environmental levels. Food wastes, when disposed of in landfills, lead to greenhouse gas emissions through the production of unregulated release of methane and through food material production activities, including the manufacture, refining, production, transport, storage, and distribution of foodstuffs. Within the general definition of global food security, the social effect of food waste may be related to a noble and moral dimension, whereas economic effects are the product of food waste costs and their impact on farmer and consumer economies. (Gustavsson et al., 2011). Various initiatives have been taken to reduce the food waste so as to help people who suffer from hunger mainly associated with poverty and poorly developed food systems. One of the sustainable development goals of the United Nations is to reduce per capita food waste by half by 2030 for global nutritional security and environmental sustainability (United Nations, 2015).

Food and agricultural waste generated during all stages from farm to fork are valuable resources which have the ability to be transformed into beneficial ones. Conversion of waste into beneficial products which may be further used worthily can be deemed as process of begetting wealth. This concept of waste management (WM) articulates the term "*waste to wealth.*" Conviction of waste to wealth is to view waste as a valuable resource which forefend it from disposal and hence preventing the environmental menace. This concept ends approaches of waste from agricultural farms and rather considered as an opportunity to create value. Considering the amount of agricultural waste generated, various innovative waste conversion processes can be used as business opportunities and a source of economic generation (Ezejiofor et al., 2014).

3.2 WASTE MANAGEMENT HIERARCHY

Waste Management Hierarchy was brought into in European policy in 1970s, with the 1975 Regulations on Waste (European Parliament Council, 1975) and then it was comprehensibly explained in European legislation in the Community Strategy for WM in 1989 (European Parliament Council, 1989). This framework is based on the '3Rs' postulation, meaning reduce, reuse, and recycle. The WM hierarchy is the set of priorities for waste reduction and management steps to be taken and is diagrammatically described as a reverse triangle (Figure 3.1). The waste hierarchy focuses on determining the solutions most likely to produce the best overall environmental performance. Options that are provided in the waste hierarchy are often chosen to provide the best environmental overcome. As shown in Figure 3.1, avoidance of waste generation is the most appropriate choice and disposal is the least beneficial option at the lowest of an inverted pyramid. Environmental conservation body with respect to food WM (EPA, 2014) defined WM hierarchy as: source reduction, feeding of hungry people, animal feeding, industrial uses, composting, and last option to be available as incineration or landfilling.

FIGURE 3.1 Waste management hierarchy (Papargyropoulou et al., 2014).

The first step involved in reducing food waste should start by addressing the unwanted food surfeit and preventing over-supply as well as overproduction of food (Papargyropoulou et al., 2014). The succeeding steps in the WM ladder involves exploitation of food waste for energy production in various industrial sectors with the aid of anaerobic digestion (AD) to produce bio-hydrogen, bio-methane etc. or the recovery of valuable phytochemicals such as bioactives, which are beneficial to food, pharmaceutical or cosmetic industry. Waste recycling can be used metonymy for energy recovery and includes collection of discarded agricultural waste like peel, pomace, seeds from processing industries, husks from farms, poultry droppings, biomass, etc., processing, and finally converting into new valuable products. The ethos of waste recycling is to minimize the quantity of waste being exposed to the environment and cause pollution which will imperil human health and sustainable development (Ezejiofor et al., 2014). Composting can also be manipulated to handle food waste or byproducts from industrial sectors to recover nutrients or development of humic substances used in the carbon sequestration process. The last and least desirable option available is landfilling or incineration. Landfilling creates environmental problem because of the fact that biodegradable organic material produces odors, causes contamination of surface water by formation of leachates, responsible for global climate changes, etc. Whereas, thermal treatment providing energy recovery from waste, is limited because of low heating values of food waste as it contains organic substances (Panda et al., 2016).

3.3 WASTE TO WEALTH STRATEGIES

The waste-to-wealth strategies can be divided into traditional or conventional methods and novel or emerging valorization methods (Figure 3.2). In the case of the conventional method, there is biological or chemical conversion of organic matter of food waste into simpler molecules such as CO_2, CO, CH_4, H_2, H_2O, NH_4, etc., and a solid inert matter like sludge, compost or ashes. Despite being able to provide useful products in conventional methods such as bio-gas, heat, bio-hydrogen, and power, these methods do not facilitate to retrieve the beneficial compounds from food and agri-industrial waste. At the same time, these strategies have a negative effect on the environment, including air pollution, emissions of greenhouse gases (GHGs), wastewater, and the processing of solid waste. However, valorization approaches concentrate on isolating useful waste components and transforming them into value-added chemicals such as antioxidants, natural colorants, dietary fibers, etc. Various valorization approaches can be used collectively to ensure besides the recovery of precious components but also the remaining waste is completely treated (Esparza et al., 2020) These techniques are preferred over the conventional strategies as they reduce the amount of waste to be disposed of while conventional methods only aim for finding solution to dispose the waste already generated so as to minimize its effect on humans health and its environment. Valorization therefore is the more coherent approach; however, the focus should be given for their economic and natural resource expenditure corresponding with them to ensure their sustainability and profitability.

FIGURE 3.2 Waste to wealth strategies (Modified from Esparza et al., 2020).

3.4 CONVENTIONAL METHODS FOR RECYCLING AND REUSING OF FOOD AND AGRICULTURAL WASTE

3.4.1 LAND FILLING

Landfilling is one of the common and simple disposal options for solid waste despite its environmental impact. Nowadays, in European countries, landfilling of organic waste has been reduced because this is anticipated to reduce methane gas emission and decrease the load on global warming from landfills (Manfredi et al., 2010). Methane produced in landfills due to AD of organic matter is considered the third-largest human-induced source of atmospheric methane emission, producing almost 800 million tons of CO_2 equivalent (Esparza et al., 2020). Food industries generate large amount of waste during various stages like processing, packaging, transportation, and storage. These wastes along with high biodegradability also possess higher chemical as well as biological oxygen demand (BOD). Leading to seasonal fluctuations and handling methods, there are also variations in composition and pH. Due to the higher amount of water and greater rate of accumulation, these wastes result in the bacterial contamination, making landfilling unsuitable for food and agro-industrial waste (Ravindran et al., 2016).

In landfills, macronutrients of waste are hydrolyzed into soluble components and consequently to the biogas via the process of methogenesis. This biogas consists predominantly of a mixture of methane and CO_2, is additionally refined to acquire bio-methane by separating out CO_2 and other secondary components (Kibler et al., 2018). This biogas can be best utilized in the generation of electricity or used in engine generators to produce both heat and power. The issue of leachate formation caused by rainwater infiltration and even the moisture contained in waste are also associated with landfills. This water, when it gets strained through the different layers of waste, gets accumulated at the bottom of the landfill, which in turn pollutes the underground table water with various organic or inorganic impurities (Bhatt et al., 2017).

3.4.2 ANIMAL FEEDING

One of the emerging subsectors of agriculture in developing countries is livestock production. However, these countries are also faced with the challenge of feed deficit. To tackle this challenge, utilization of agriculture waste is considered as the best possible alternative. These wastes are equipped with both economical as well as nutritional advantages, providing desirable nutrients and minimizing the cost of feeding (Wadhwa et al., 2013). However, the usage of agricultural by-products as a feed supply has its own drawbacks and dangers, such as the inclusion of some harmful elements in it, contributing to disease spread or the failure of certain animals to take away all food or residues. Proper and careful observation is also necessary prior to animal feeding. Nevertheless, when used as a partial feed substitute, livestock may be helpful. Transport costs for waste collection, costs associated with thermal treatments to make it suitable for animal use, sometimes render this alternative unfeasible. Compared to other conventional methods like landfilling or AD, animal feeding is advantageous when used under controlled conditions (Salemdeeb et al., 2017).

Various countries manipulate agricultural waste as animal feeding viz. Japan and South Korea. Percentages of food waste reused as feed are 35.9% and 42.5%, respectively. However, in the European Union, it is considered illegal to use animal feed made from recycling of the food waste (Salemdeeb et al., 2017).

3.4.3 COMPOSITING AND VERMI-COMPOSITING

Compositing is an aerobic process where organic matter of waste is converted into material possessing smaller molecular chains, high stability and humus-rich properties. In the aerobic world, the processes are carried out using many organisms, such as bacteria, algae, fungi, protozoa, actinomycetes, which are naturally present or artificially introduced to the biomass. The resulting material has got a tremendous potential to be used in agricultural crops and its utilization in recycling of organic matter of soil (Shilev et al., 2007).

Vermicomposting, on the other hand, is the process of conversion of biodegradable waste material into good quality manure by use of earthworms specifically *Eisenia foetida species*. This method is economical and environment friendly for producing organic matter and becomes a valid method of bio WM. In vermi-compositing stabilization of organic matter occurs by combine effect of earthworms and aerobic microorganisms. The process of vermi-composting results in an overall decline in total organic carbon and carbon-nitrogen (C/N) ratio, however, it manages to increase the content of nitrogen-phosphorus-potassium (NPK) relative to the unprocessed compost and other agricultural wastes (Gupta et al., 2019).

Compositing and vermicomposting of food and agricultural waste have environmental benefits compared to landfilling as lower greenhouse gas emission and leachate formation. However, it needs various requisites to be fulfilled such as sorting of organic fraction of waste, energy requirement for the process of aeration and mixing, addition of water and finally controlling various processing parameters like temperature, humidity, etc. Under inappropriate operating conditions, odor emissions would take place, paving a threat to environment together with the creation of cheap quality compost (Cerda et al., 2018).

3.5 ADVANCED VALORIZATION METHODS FOR RECYCLING AND REUSING OF FOOD AND AGRICULTURAL WASTE

3.5.1 BIOACTIVE COMPOUNDS FROM FRUIT AND VEGETABLE WASTE

Fruit and vegetable by-products are ample sources of bioactives like phenolic compounds, dietary fibers, and other phytochemical compounds and have been examined for the extraction of such compounds from it (Galanakis, 2012). These phytochemicals associate with proteins, fatty acids and other biological molecules to synthesis valuable effects to human health in respect of health upgradation and risk of disease reduction (Kumar et al., 2017). In many fruits and vegetables, pulp or flesh is consumed while seeds, peel, core or pomace produced in processing industries is mostly thrown as waste, however, experimental studies have shown that notable amounts of valuable chemicals and essential nutrients are found in these parts of fruit such as the lemon peel, grapes skin, and orange peel, and the seeds of longans, avocados, jackfruits, and mango peel possess as high as 15% phenolic concentrations than the pulp of fruit (Soong et al., 2004).

Functional composition of agricultural waste is attracting great interest because of the shift of diet from animal origin foods to plant-based foods. Structural and functional features of bioactive compounds extracted from fruit and vegetable waste (FVW) range greatly, such as dietary fiber, terpenes, phenolic compounds, fatty acids, saponins, phytoestrogens, etc. Such substances have considerable potential for utilization in the dairy, medicinal, cosmetic, and garment industries. (Esparza et al., 2020) Interestingly, in the last few years, various bioactive compounds from FVW have been sold as nutraceuticals or dietary supplements in the last few years (Table 3.1).

3.5.2 PHENOLIC COMPOUNDS

Plants synthesize phenol compounds as secondary metabolites which are not used for their daily functioning; however, they are responsible for the sensory and nutritional of horticultural products along with other functions. Polyphenol consists of one aromatic ring or rings with either hydroxyl group or groups. Structurally, polyphenols may cover from plain phenol to the highly polymerized compounds (Balasundram et al., 2006). These polyphenolic compounds can be grouped into various classes based on the number of phenol rings present within the compound and also the structural element that binds these aromatic benzene rings. These include flavonols, flavanones, flavones, flavanonols, anthocyanidins, isoflavones, flavonols, tannins, stilbenes, phenolic acids, and lignans. (Robbins et al., 2003), included other classes.

The peel, seeds, and pomace of horticultural products mainly considered as waste contain high amounts of phenolic compounds

TABLE 3.1   Patented Methodologies Leading to Commercial Applications of Product-Specific Food Wastes (Galanakis et al., 2012)

Food By Product Source	Applicant	Title	Products	Applications	References
Citrus peel by-product	Tropicana products Inc. (Florida, USA)	Treatment of citrus fruit peel	Sugar syrup	Naural food sweetener	Bonnell (1983)
Cheese whey	Alta-Lava Food Engineering AB (Lund, Sweden)	Method for obtaining high-quality protein products from whey	α-lactoalbumin and β-lactoalbumin containing products	Food supplement and additive	Jensen and Laren (1993)
Olive mill by-product	Davisco International Foods Inc. (Le Sueur, USA)	Isolation of glycoprotein from bovine milk	Whey protein isolate/ Bipro®	Food supplements	Davis et al. (2002)
	Kraft Foods Holding Inc. (Northfield, USA)	Method of deflavoring whey protein using membrane electrodialysis	De-flavored whey proteins	Food supplements	Crowely and Brown (2007)
	Cre Agri, Inc. (Hayward, USA)	Method of obtaining a hydroxytyrosol-rich composition from vegetation water	Hydroxytyrosol/Hidrox	Food supplements	Crea (2002)
	Consejo Superior de Investigaciones Cientificas (Madrid, Spain)	Method for obtaining purified hydroxytyrosol from products and by-products derived from olive tree	Hydroxytyrosol (99.5%) Hytolive®	Conserving foods, functional ingredients in bread	Fernandez-Bolanos et al. (2002)
	Phenoliv AB (Lund, Sweden)	Olive waste recovery	Olive phenols and dietary fibers containing powders	Natural antioxidants in foods	Tomberg and Galanakis (2008)
Tomato by-product	Biolyco SRL (Lecce, Itlay)	Process for the extraction of lycopene	Lycopene	Food antioxidants and supplements	Lavecchia and Zuorro (2008)

TABLE 3.1 *(Continued)*

Food By Product Source	Applicant	Title	Products	Applications	References
Soy protein isolate wastewater	Shang Dong Wonderful Industry Group Co., Ltd. (Shangdong, China)	Method for extracting and recycling albumin from whey wastewater from production of soy protein isolate	Soybean albumin	Food additives and supplements	Jinshan et al. (2009)
Shrimp and crab shell	Qingdao Zhengzhongjiahe Export and Import Co. Ltd. (Shandong, China)	Preparation of chitosan derivative fruit and vegetable anti-staling agent	Chitosan ($\geq$85%) food grade	Food thickeners	Shenghui (1995)
De pectinated apple pomace	Yantai Andre Pectin Co. Ltd. (Yantai, China)	Process for extracting non-pectin soluble pomace dietary fibers	Apple dietary fiber granules	Dietary supplement	Anming et al. (2010)
Pomegranate rind and seedcase residues	Xi an app chem-bio (Tech) Co., Ltd.	Method for preparing punicalagin and ellagic	Ellagic acid (40%) and punicalagin (40%)	Food antioxidants and cosmetics	Guangyu and Zhang (2011)

like. It was reported that potato peel constitutes 50% of phenolic compounds out of total bioactive components (Friedman et al., 1997). Choi et al. (2016) reported that the "Superior" variety of Korean potato peel possess higher quantity of phenols viz. chlorogenic acid 385 ± 50 µg/g DW, 21.9 ± 2.0 µg/g DW chlorogenic acid isomer II, and 103 ± 10 µg/g DW caffeic acid than the pulp such as 107 ± 4 µg/g DW chlorogenic acid, 4.2 ± 1.2 µg/g DW chlorogenic acid isomer II, and 2.29 ± 0.51 µg/g DW caffeic acid of the same variety. Studies have shown that citrus by-product is an ample provenance of phenolic compounds because, relative to the inner edible one of the fruit, it contains larger amounts of polyphenols (Balasundram et al., 2006). Not only citrus peel, but peel of various fruits like apple, peach, and pear contain higher amount of phenolics than pulp (Gorinstein et al., 2001) yield of phenolic compounds from horticultural waste depends on the method used for its extraction, e.g., compared to the maceration process, the phenolics from kinnow peel were increased by four times using ultrasound-assisted extraction (UAE) (Safdar et al., 2016).

3.5.3 DIETARY FIBER

Dietary fiber is another valuable phytochemical used in various fields in particular milk, health care, and polymer industries. For these industries to develop economically, the recovery of dietary fiber from agricultural waste is an important raw material. A composite mixture of plant-based non-digestible polysaccharides, waxes, and lignin is reflected in the word "dietary fiber." Cellulose, hemicellulose, and lignin are insoluble fibers, while pectin, fructans, arabinoxylans, and β-glucans are soluble fibers. The physiological behavior and dietary advantages of these two separate fractions differ. However, soluble fiber tends to control blood glucose levels and reduce blood cholesterol, primarily used as a laxative to improve bowel movements and often to boost the development of intestinal microflora, along with probiotic organisms. These dietary fibers also play an important role in the food sector and have been used as agents for gelling, thickening, and stabilizing (Barrera et al., 2002). Various innovative functions of dietary fiber such as pectin have been found like fat replacer in the meat industry (Zhang et al., 2018). Dietary fibers have often been used in recent research as a wall material for bioactive compounds by encapsulation technology (Mueller et al., 2018) or as a delivery substance for antimicrobial and antioxidant compounds in the edible packaging method.

Various fruit and vegetables waste have been investigated as a source of dietary fibers such as grape pomace was evaluated to be a bountiful in dietary fibers, viz, cellulose, hemicelluloses, and small amount of pectins (Kammerer et al., 2005). Around 10 different varieties of grapes were investigated for dietary fiber content by (Gonzalez-Centeno et al., 2010). They found that red grape variety "Tempranillo" had dietary fiber in the pomace (36.90 g/100 g fruit waste, stem 34.80 g/100 g FW) while whole fruit contained 5.10 g/100 g food waste. Carrot pomace was also found to be good source of dietary fiber, and it was explored to possess about 63.6% dry matter, of which insoluble one was 50.10% dry matter and soluble fibers were 13.50% dry matter (Chau et al., 2004).

3.5.4 ENZYMES

Enzymes are essentially proteins that, along with an amino acid sequence, require either a co-factor in the form of inorganic ions or a co-enzyme for their activity as organic or metallo-organic molecules. In separate biochemical reactions, enzymes serve as biological catalysts (Sharma et al., 2016). Enzymes have a predominant function in food and pharmaceutical industry. Uses of enzymes is also extended to the valorization of FVW for product development, energy recovery, etc. Commercial enzymes are usually very expensive because of their manufacturing procedure and raw materials cost. For lowering down the price of enzyme synthesis, utilization of agricultural waste as a substrate for their production proves to be an efficient strategy. Also a wide variety of enzymes can be obtained by using FVW as their substrate (Ravindran et al., 2016). The use of enzymes is extended to various industries such as amylase and pectinase in food processing industries, tannase to reduce tannic acid concentration in tannery effluent and cellulases in bio-fuel industries.

Solid-state and submerged fermentation (SSF) are the two bioprocesses utilized for enzyme production. SSF can be described as a process where microbes are grown on solid materials created from food and agro-industrial waste in the absence of free liquid. While SMF can be explained as a process in which the growth of microbes is done in a substrate with free-flowing water (Kapoor et al., 2017). Soluble portions of substrates are

94 Integrated Waste Management Approaches for Food and Agricultural Byproducts

dissolved in the liquid phase in this procedure, while insoluble substrates are suspended or submerged. This fermentation approach allows for easy regulation of the reaction parameters, but efficiency is poor and energy requirements are high. FVW is best suited for SSF and provides microorganisms with equal habitat to their natural ecosystem. This method favors attaining elevated product yield and cell density compared to the SMF. However, disadvantages of solid-state fermentation (SSF) are linked to complex process control and scale-up.

3.5.5 AMYLASES

Amylases are a group of enzymes which comprises of three different types, namely α-amylase, β-amylase, and GA. These amylolytic enzymes act on polysaccharides and other oligosaccharides and convert into simple, possessing lower molecular weight than polysaccharides viz glucose, fructose, maltose, etc. The starch molecule consists of amylose composed structurally of a linear chain of alpha 1,4 bonded unbranched D-glucose units and amylopectin with firmly branched alpha 1–6 bonded D-glucose residues. Amylases can be distinguished according to the site of action of hydrolysis into exo and endo-amylase. Exo-amylase hydrolysis α-1,4 and some exo-amylases like GA attack both alpha-1–4 and alpha-1–6 bonds in order to produce simpler sugars. In amylopectin and related complex polysaccharides, endo-amylase attacks α-1,4 starch bonds and does not interact with α-1,6. Loquat (Eriobotrya japonica Lindley) kernels were used to generate alpha-amylase in SSF using Penicillium expanse, a waste extracted from loquat fruit. In optimized conditions with a temperature of 30°C, moisture content of 70%, particle size 1 millimeter, pH 6.0 and starch and peptone were 1,012 U/g of loquat kernel flour as supplements of the enzyme produced. The mango kernel was also used as an amylase processing substrate and 0.889 U/g of the mango kernel was the maximum enzyme produced. Similarly, banana waste was also used for the synthesis of α-amylase using *Bacillus subtilis*. Various microorganisms are employed for the production of amylase such as *A. niger, Aspergillus awamori, Rhizopus oryzae, Aspergillus oryzae, Bacillus subtilis, Bacillus licheniformis, Candida guilliermondii, Aspergillus tamarii,* and *Thermomyces lanuginosus*. Commercially, *Bacillus subtilis, Aspergillus niger,* and *Rhizopus oryzae* are the most used species in various industries (Said et al., 2014).

3.5.6 PECTINASES

Pectinases are the group of enzymes that act on pectic substances and are important component in fruit and vegetable cell walls. These pectinases are classified into two major groups: depolymerizing and demethylating enzymes. Depolymerizing breaks α-1–4 linkages in principle pectin chain like galacturonase and pectin lyase while as demethylating enzymes esterifies pectin to pectic acid by eliminating methoxy residues such as pectinesterase (Mrudula et al., 2011). Pectinases are used in the degradation of plant material hence is mostly used for facilitating juice withdrawal from fruits. Various agricultural wastes have been utilized for the production of pectinases using SSF such as grape pomace has been explored as a substrate for the production of pectinases using *A. awamori* yeast (Botella et al., 2005). Mrudula et al. (2011) used six separate orange peel, lemon peel, banana peel, rice bran, wheat bran, and sugar cane bagasse horticultural waste for the processing of pectinase using *A. niger* from the SSF. Among them, orange peel had the highest value of pectinase production amongst all substrates (1,224 U/g of dry substrate).

Pectinases have important applications in the food industry, mainly in fruit juices and wine industry for withdrawal, clarification, and concentrating of juice from their source. Also, these have been used in the withdrawal of flavors, color pigments, and essential oils from their plant residue (Castilho et al., 2000).

3.5.7 INVERTASE

Invertase enzyme (β-fructofuranosidase) is being used for invert sugar production. Invert sugar has low crystallinity collated to sucrose even in high concentration, thus retaining the products in which invert sugar is being used fresh and soft for an extended period of time. Fruit peel combinations (sapota, pineapples, and bananas) were used as substrate for invertase production. It has been found that invertase produced from fruit peel waste was having lower activity than fructose as the important carbon source while as have higher activity than lactose as a carbon source (Mehta et al., 2014). Uma et al. (2010) also explored the invertase synthesis by *Aspergillus flavus* exploring waste like fruit peel waste as a fermentation medium. They evaluated that under optimized conditions such as temperature 30°C, inoculum 3%, pH 5 and incubation period of 4 days, high level of invertase was produced. They also observed that the addition of yeast extract and sucrose magnified the synthesis of invertase. Invertase is mostly used in food industry for manufacturing of candies, jam, confectionery, and also used in pharmaceutical industry for production of syrups, etc. (Panda et al., 2016).

3.5.8 TANNASE

Tannase also known as Tannin acyl hydrolase, is used to produce gallic acid and glucose by hydrolyzing tannin. Tannins are present in plant-based materials in abundance. Mostly fungi are explored for production of tannase, however, some bacterial species namely Bacillus sp. and Lactobacillus sp. are being evaluated for tannase producing activity (Panda et al., 2016). Various agro-industrial wastes as substrate have been demonstrated for tannase production. Cashew apple bagasse using A. oryzae was explored as a substrate for the tannase production. It was observed that the highest yield of 3.42 U/g at 24 hours and 0.128 U/g was procured at 48 hour fermentation. They also presented that supplementation of media with 2.5% ammonium sulfate enhanced the productivity of tannase (Prommajak et al., 2014). Grape peel was also used as a base for the production of tannase utilizing various fungal strain co-cultures: A. niger, P. chrysogenum, and T. viridae. The results showed that the highest enzyme activity of 84 U/g/min was obtained at an incubation period of 96 h in combination of P. chrysogenum and T. viridae (Paranthaman et al., 2009). Tannase is utilized in the food and beverage industry to suppress chemical astringency, which is mostly due to tannins and also effective in reducing the accumulation of tannic acid in tannery effluent.

3.5.9 ORGANIC ACIDS

Organic acids are distinguished by poor acid properties and are known to be elementary units for microbial processing chemicals. Lactic acid, acetic acid and citric acid, are the widely consumed organic acids. Commercially, bacterial, and fungal species are mainly utilized for the synthesis of organic acids. Bacteria such as *Bacillus sp., Lactobacillus sp., Arthrobacter paraffinensis thermophillus, Arthrobacter paraffinensis,* and *fungus* such as *Aspergillus sp., Yarrowia lipolytica., Penicillium sp* and various yeast species are used to synthesize organic acids. Different industries use organic acids, such as food, nutrition, pharmaceutical industry, oil, and gas units, etc. FVW have been utilized as substrate for the bioprocessing of numbers of organic acids (Panda et al., 2016).

Citric acid is a carboxylic acid generally present in biological organisms which include animals, plants, and microorganisms. It is available as middle compound of tricarboxylic acid cycle. *Aspergillus niger* with apple pomace as substrate have been used to produce up to 80% citric acid (Dhillion et al., 2011). Kareem et al. (2011) experimented on the synthesis of nitric acid using *Aspergillus Niger* with banana peel as a substrate for fermentation. The acetic acid production was about 82 g citric acid/kg dry wt. They further checked that the key substrate production of citric acid was escalated with the inclusion of trace elements and other nutrients in the substrate medium with dried banana peel. The citric acid yield from the banana peel was around 90% of the dry weight of the batch. Imandi et al. (2008) also synthesized citric acid from pineapple waste used as substrate for fermentation using the yeast *Yarrowia lipolytica.* Citric acid has an important role in food and pharmaceutical sector as natural preservative.

Acetic acid is another organic acid which is consumed mostly as vinegar. Percentage of acetic acid in vinegar varies in different countries usually (2.5–11%). Acetic acid is produced by aerobic fermentation as well as anaerobic fermentation. However, *Clostridium* and *Acetobacterium* are able to produce acetic acid in a single step by anaerobic fermentation of glucose although, food-grade acetic acid is mainly generated via oxidative fermentation by acetobacter (Panda et al., 2016). Various successful extensive research studies have been brought to establish acetic acid production from fruit wastes. Pineapple peel has been used as a fermentation substrate for acetic acid production. The peel was first anaerobically fermented for 48 hours using *Saccharomyces cerevisiae* for conversion of sugars present in peel into the ethanol, after that *Acetobacter aceti* was used for further conversion of ethanol into acetic acid using about 9 days incubation period. It has been found that maximum acetic acid was 4.77% at optimal conditions (Raji et al., 2012). Vikas et al. (2014) also evaluated papaya peel as fermentation media for producing acetic acid. Papaya peel hydrolyzates were first manufactured and then anaerobic fermentation was accomplished by *Acetobacter aceti* for ethanol content production (8.11%) and then into acetic acid. At the end of the fermentation process, the titratable acidity of the final broth processed acetic acid was 5.23%.

The most commonly used organic acid worldwide, lactic acid, is produced both chemically and by the biological fermentation process. About 90% world's lactic acid generation is bacterial fermentation and produces nearly 90% purity lactic acid (Vijayakumar et al., 2008). Lactic acid plays a vital importance in several key industries including food, chemical, pharmaceutical, cosmetic, etc., and acts mainly as acidulant and preservative. Also, lactic acid has been conferred with GRAS status for human consumption. Biodegradable polylactic acid and biocompatible polylactate polymers are formed by the lactic acid polymerization reaction. Generation of lactic acid from agricultural by-products have been well cited in the literature. Mudaliyar et al. (2012) explored a study on peels of different fruit (mango, orange, and sweet corn) and vegetables (potato, green peas) for the generation of lactic acid with the help of *Lactobacillus casei* and *L. delbruckii* via fermentation process. The authors concluded that the maximum quantity (63.33 g/L with *L. casei*) of mango peel was made, followed by orange peel (54.54 g/L with *L. delbruckii*), potato (38.88 g/L with *L. casei*) and sweet maize (37.62 g/L with *L. casei*). However, orange-peel *L. delbruckii* showed a poor output among other wastes.

3.5.10 EXOPOLYSACCHARIDES

Exopolysaccharides are those biopolymers which are synthesized by microorganisms extracellularly or secreted into the extracellular medium throughout their growth. Utilization of agricultural waste as a medium for synthesis of biopolymers will fathom the issues of WM and as well as reduce the cost of biosynthesis of valuable metabolites. Microorganisms synthesize an ample variety of such biopolymers existing with distinctive chemical and physical properties. These exopolysaccharides have an advantage over the plant-derived or synthetic polysaccharides as they are nontoxic, resistant to mechanical and oxidizing agents, biodegradable, resistant to pH and temperature values (Pirog et al., 2016). Based on the structure and composition, they are categorized into two subgroups viz Homo and heteropolysaccharides. Homopolysaccharides are those which possess identical monomeric units which can be either D-glucose or D-fructose. Thus depending on their monosaccharide units, they can further be divided into glucans and fructans. Heteropolysaccharides are constructed primarily by glucose, galactose, and rhamnose in different ratios (de Vuyst et al., 2001).

Large scale fermenters with submerged cultures are used for commercial production of exopolysaccharides where process parameters are carefully monitored and controlled. This process is principally favored by the occurrence of excess carbon source, accompanying with constraint by others such as oxygen or nitrogen (Kumar et al., 2007). Although synthesis of these exopolysaccharides is quite extravagant as the feedstock used in industrial production are mainly glucose and fructose. Hence, there has been forage for alternative and cheaper carbon source for the production of these biopolymers. Horticultural waste has been demonstrated as an efficient and cost-effective carbon source for the production of exopolysaccharides. Table 3.2 shows various agricultural wastes as carbon source for the synthesis of different exopolysaccharides and their industrial application. Xanthan gum is a heteropolysaccharide recommended by FDA in 1969 as the first biopolymer to be used as a food additive. Xanthan gum is synthesized primarily from *Xanthomonas campestris* as a part of their metabolism. The main chain consists of glucose residues with trisaccharide side chains containing glucuronic acid, mannose, pyruvil, and acetyl residues (Kanimozhi et al., 2018). Due to the unique rheological properties of xanthan gum like pseudo-plasticity, high viscosity even at low concentrations and stability over ample extent of pH and temperature, this biopolymer find many practical approaches in various fields like food, cosmetic, pharmaceutical, textile, etc. In the food industry, Xanthan gum can be utilized as gelling, stabilizing, thickening, and suspending agent in fruit juices, sauces, syrups, creams, ice-creams, and deserts (Hublik, 2016).

100 Integrated Waste Management Approaches for Food and Agricultural Byproducts

TABLE 3.2 Different Waste Sources for Synthesis of Exopolysaccharides and Their Potential Uses

Exopolysaccharide	Agricultural Waste Source	Applications
Xanthan (*Xanthomonas campestris*)	Date palm juice by-product, coconut shell, cocoa husk, cassava serum, sugar beet molasses, sugar cane broth, broom corn stem	Food industry, pharmaceutical industry, cosmetics, paints, textile printing, agricultural products
Curdlan (*Agrobacterium species*)	Orange waste, asparagus waste, date palm juice, cassava starch hydrolysates, coconut water, sugarcane molasses	Food industry as fat mimic, films, texturizer, biomedical industry as immunostimulator, bioabsorbent for heavy metal remover.
Pullulan (*Aureobasidium Pullulan*)	Coconut by-products, cassava bagasse, rice hull, potato starch	Food industry as food coatings, low calorie ingredient, prebiotic, pharmaceutical industry as drug delivery, tissue engineering, vaccination, oral care.

Curdlan is another exopolysaccharides which is linear, water-insoluble composed primarily of β (1–3)-linked glucose produced by fermentation with Agrobacterium species. This name was given because it curdles when being heated. This ability makes it a good gelling property to enhance textural properties, water holding capacity and thermal stability of commercial products. Curdlan belong to the β-glucans family of biopolymers. Depending on heating temperature, this exopolysaccharides has the capability of producing gels of two types: thermo-reversible and thermo-irreversible hydrogels. When heated to temperatures of 50–60°C and cooled, weak gel is created, that returns to liquid state upon reheating. But after heating to temperature above 80°C, high set thermo irreversible gel is formed (Zhang et al., 2002). These distinctive characteristics of exopolysaccharide have grabbed the attention of the food industry. In food industry, it has capability of being used as fat mimic in meat products like sausages (Funami et al., 2006), improve textural properties of starch noodles, development of stable and functional yogurt, used as component in edible and biodegradable films, etc. Nowadays, it has also been used in biomedical applications as immune-stimulator, also acts adsorbent for various heavy metals.

Pullulan is another developing biopolymer that has a beneficial function in the pharmaceutical and food sectors, serving as a source of polymeric materials. This exopolysaccharides is synthesized when fermentation is done using yeasts such as *Aureobasidium pullulans* (Prajapati et al., 2013). Pullulan has important functions in the food industry and can be used in food coatings as biodegradable, edible, and oxygen impermeable films. It can also be used as prebiotic, emulsifier, low calorie ingredient and as partial replacer of starch in pasta (Farris et al., 2012). Various carbohydrate sources has been incorporated into the fermentation media for production of Pullulan. Agro-industrial waste has been used as media for its production, and the results have revealed that higher or same yield of pullulan can be obtained when collated to conventional substrate. Uses of agro-industrial waste as media can be advantageous and also economically sound as these substrates are cheap and environmentally sustainable (Prajapati et al., 2013).

3.6 NANOPARTICLE SYNTHESIS

In contrast to traditional processing protocols containing hazardous chemicals and solvents, nanoparticles biosynthesized by the use of horticultural waste have produced sustainable, efficient, and ecologically sound technologies with a lesser peril to human health and the environment. There are some useful bioactive molecules in FVW, like polyphenols, alkaloids, amino acids, enzymes, proteins, polysaccharides, tannins, saponins, vitamins, and terpenoids and other compounds that play an active role as reducing agents in the production of metal nanoparticles. Some of these biomolecules serve as a modeling agent that drives particle development in a particular direction, while some others act as a capping agent to avoid agglomeration of nanoparticles (Kumar et al., 2020). The important characteristic of nanoparticles are immensely smaller size and large surface area to volume ratio that find its application in different sectors such as pharmaceuticals, food, and biomedicine. Such as gold (Au), ZnO, copper oxide (Cu_2O, CuO) nanoparticles are widely used in medicine, food packaging industry, etc. Au nanoparticles find their application for diagnostics, targeted delivery of medicines and tumor destruction through hyperthermia, metal oxide nanoparticles like copper oxide and zinc oxide (ZnO) have shown antimicrobial activity (Ghosh et al., 2017).

Patra et al. (2016) used dried onion peel extract to synthesize gold nanoparticles by reducing Au^{3+} to onion peel-gold nanoparticles by forming colloidal solution. The biomolecules present in onion bulb mainly the cysteine derivatives might be responsible for nano-particles synthesis. Potato peel has also been used for zinc oxide nanoparticles synthesis as starch of potato peel aids in reduction of metal ion (Bhuvaneswari et al., 2017). Various fruit peel extracts such as banana, lemon, and pomegranate has been used for silver nanoparticles synthesis in which silver ions in aqueous solution has been reduced into silver nanoparticles (Skiba et al., 2019). Bottom-up approach is being used for synthesis of nanoparticles. Atoms and molecules unite during the process to create precursor building blocks that ultimately self-assemble to form nano-range particles. Biogenic reduction proceeds promptly, and as this process goes on, there is a characteristic change in color of the solution, suggesting nano-particles synthesis as shown in Figure 3.3, such as during silver nanoparticles synthesis using reducing agent as orange peel extract, the color change in reaction mixture from colorless to yellowish-brown indicates nanoparticle synthesis (Ghosh et al., 2017).

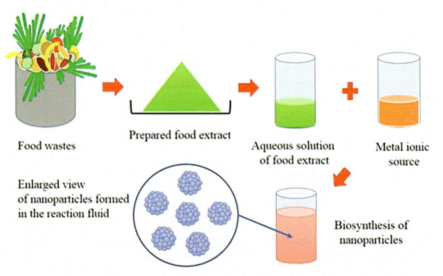

FIGURE 3.3 Schematic representation of the biogenic synthesis of nanoparticles using horticultural food waste extracts. (Reprinted from Ghosh et al., 2017. https://creativecommons.org/licenses/by/4.0/).

3.7 BIOPLASTICS

Plastics are an important category of polymers possessing many applications due to its supreme properties as mechanical resistance, lightness, versatility, low cost and easy processability. Yet, at the same time, under natural conditions, plastics have a very low degradation rate, so they are known to be chronic environmental contaminants. In addition, they are made from derivatives of petroleum, which are the cause of high carbon footprints. For such problems, there is a huge need in evolving bioplastics and that too from renewable resources (Esparza et al., 2020). Bioplastic production using food waste can be considered as renewable sustainable process as the material used for their synthesis is neutral carbon sources. This method will help in competitiveness and expansion of bioplastic industries by reducing the raw material cost. These bioplastics are mainly composed of polyesters. The various biodegradable polyesters are poly(hydroxyalkanoates) (PHAs), polylactic acid (PLA), polybutylene succinate (PBS), polyglycolic acid (PGA), polybutylene succinate adipate (PBSA), poly ε-caprolactone (PCL), aliphatic-aromatic copolyesters (AAC), polybutylene adipate/terephthalate (PBAT), and polymethylene adipate/terephthalate (PTMAT) (Gonçalves de Moura et al., 2017). Among them, main bioplastics are polylactic acid and polyhydroxyalkanoates. PLA synthesized through the polymerization of lactic acid (2-hydroxy propionic acid) is a linear aliphatic polyester. It is produced both from both fossil and renewable resources and is considered as biodegradable thermoplastic polyester. It is one of the promising polymer which is being used commercially as replacer of various plastics such as low density polyethylene (LDPE), high density polyethylene (HDPE), polyethylene terephthalate (PET), and polystyrene (PS) (de Moura et al., 2017). Renewable agricultural sources, like corn, or other carbohydrate sources like simpler sugar dextrose and then by fermentation process into lactic acid can be used for synthesis of PLA. PLA certified as generally recognized as safe (GRAS) according to the United States Food and Drug Administration is biocompatible as well as biodegradable and Farah et al. (2016). PLA finds its application in various sectors such as in industrial packaging and biomedical applications.

Polyhydroxyalkanoates (PHA) are polyester bioplastics produced mostly by heterotrophic microorganisms by polymerization of R-hydroxyalkanoic acids. These microorganisms assemble PHA as carbon and energy storage in the cytoplasm. Thus capacity of cells to accumulate PHAs and the cell density are the limiting factors for PHA production during the fermentation process (Steinbüchel et al., 2001). Various naturally PHA producing microorganisms have been recognized while as few bacteria have been used commercially for PHA production which include *Alcaligenes latus, Bacillus megaterium, Cupriavidus necator,* and *Pseudomonas oleovorans.* These bacteria are able to convert different kinds of carbon sources into PHA. *C. necator* is one of the most widely utilized microbial strains to produce PHA commercially (Reddy et al., 2003). PHA industry utilizes food crops, vegetable oil and sugar cane as raw material for their synthesis. Amidst various sources for PHA production, agro-industrial wastes such as lignocellulosic waste from the food industry, glycerol from biodiesel production and petrochemical plastic waste have attracted attention because food waste provides strategic means of reducing the overall cost of production (Tsang et al., 2019). Used cooking oil, which is considered as waste after several uses is the most investigated food waste for PHA production as its long and medium-chain fatty acids behave as pioneer for the synthesis of various PHAs. Follonier et al. (2014) used different fruit pomace as carbon source and waste frying oil as precursor during synthesis of PHA. Food industry waste including fruit pomace (apricots, cherries, and grapes) and waste frying oil has been found to substitute the expensive sugars and fatty acids usually used during PHA synthesis as carbon feedstock for bacterial fermentations. Application of various fruit pomaces like grapes, cherries, apricot, and also the frying oil considered as waste after several uses as carbon substrate for synthesis of PHA has been well cited in literature. PHA has been explored extensively for different purposes such as packaging, medical application, energy, and fine chemicals. In the packaging industry, PHA has some unique superior physicochemical properties such as oxygen barrier better than polypropylene and polyethylene terephthalate, water vapor barrier better than polypropylene and also odor barrier. These properties have promoted PHA as packaging material in various fields like food packaging (Ahmed et al., 2018).

3.8 CARBON DOTS

Carbon dots are highly fluorescent nano-materials which belong to the carbonaceous family possessing sizes less than 10 nm. As these nano-materials were discovered recently in year 2004, but as a result of their high fluorescent property grabbed the attention of researchers. Carbon dots are of more scrutiny collated to traditional quantum dots because as they are considered to be frugal as well as environment friendly (Himaja et al., 2014). Two methods, top-down, and bottom-up, create these carbon dots. The top-down solution includes the degradation of large carbon structures in the manufacturing process by chemical oxidation using acid, electro-oxidation or laser ablation. In their synthesis, such extreme conditions and complexity are seen as one of the drawbacks of this process. Although the bottom-up solution is shown to be preferable over the top-down approach using food and agro-industrial waste, this strategy becomes chemical-free. Food and agricultural by-products as a carbon source for the synthesis of carbon dots offer remunerative benefit and become as one of the utmost captivating source of carbon for the production of CDs (Kumar et al., 2020).

Watermelon peel was used for the synthesis of carbon dots and process include mainly two processes viz. carbonization employed at 220C for 2 hours and then filtration using 0.2 μm membrane filters. The results presented that blue photoluminescence intensity was powerful. Also it has been found that photoluminescence lifetime was acceptable and stable under pH range of (2.0–11.0) and elevated salt concentration (Zhou, et al., 2017). In another study banana peel fresh and yellow in color was practiced as starting material for synthesis of carbon dots. In this study material was first heated at 200C for 2 hours in an oven for the carbonization process. The photographs of carbon dots synthesized showed blue under ultraviolet (UV) light at 365 nanometer and could also be seen by the naked eye (Vikneswaran et al., 2014).

3.9 FUTURE PROSPECTUS AND CONCLUSION

In the current situation, the evolution of a viable approach for the regulation of agricultural waste to convert into useful resource has become highly necessary and serves the primary demands of society. Such sustainable strategies should be capable of harnessing the full value of food waste and of helping to gain societal, economic, and environmental benefits. By producing value-added goods such as nanoparticles, carbon dots, vitamins, probiotics, enzymes, etc., proper use of food and agro-industrial waste that is not suitable for human consumption can be achieved. This would be an economically sound and viable means of creating innovative break-throughs in industry. Many of the methods involved in the use of food waste for useful items are at an early stage and the technological protocols and results are lacking. Therefore, with the initial investment funding, there is a very significant need to establish consortiums of researchers and industrialists to maximize the remunerative value of food and agri-industrial waste.

KEYWORDS

- **anaerobic digestion**
- **bioactive components**
- **bio-plastics**
- **nano-particles**
- **waste management**

REFERENCES

Ahmed, T., Shahid, M., Azeem, F., Rasul, I., Shah, A. A., Noman, M., Hameed, A., et al., (2018). Biodegradation of plastics: Current scenario and future prospects for environmental safety. *Environ. Sci. Pollut. R., 1–12.*

Anming, Z., De, F., Xiaoyan, H., Jianmin, Z., Bing, L., Lei, X., et al., (2010). *Process for Extracting Non-Pectin Soluble Pomace Dietary Fibers.* State intellectual property office of the people's republic of China. CN 101817809 (A).

Ariga, T., Yamazaki, E., Yamashita, K., Sasaki, M., Yamatsugu, N., & Ishii, N., (1999). *Protein Food.* Japanese Patent Office. JP 199980075070.

Balasundram, N., Sundram, K., & Samman, S., (2006). Phenolic compounds in plants and agri-industrial by-products: Antioxidant activity, occurrence, and potential uses. *Food Chemistry, 99*(1), 191–203.

Barrera, A. M., Ramırez, J. A., González-Cabriales, J. J., & Vázquez, M., (2002). Effect of pectins on the gelling properties of surimi from silver carp. *Food Hydrocolloids, 16*(5), 441–447.

Bhatt, A. H., Karanjekar, R. V., Altouqi, S., Sattler, M. L., Hossain, M. D. S., & Chen, V. P., (2017). Estimating landfill leachate BOD and COD based on rainfall, ambient temperature, and waste composition: Exploration of a MARS statistical approach. *Environ. Technol. Innovat., 8,* 1–16.

Bhuvaneswari, S., Subashini, G., & Subramaniyam, S., (2017). Green synthesis of zinc oxide nanoparticles using potato peel and degradation of textile mill effluent by photocatalytic activity. *World J. Pharm. Res.*, *6*, 774–785.

Bonnell, J. M., (1983). *Treatment of Citrus Fruit Peel.* Australian Government e IP Australia. 1130883 (A).

Botella, C., De Ory, I., Webb, C., Cantero, D., & Blandino, A., (2005). Hydrolytic enzyme production by Aspergillus awamori on grape pomace. *Biochem. Engr J.*, *26*, 100–106.

Castilho, L. R., Medronho, R. A., & Alves, T. L., (2000). Production and extraction of pectinases obtained by solid-state fermentation of agro-industrial residues with *Aspergillus Niger. Bioresour. Technol.*, *71*, 45–50.

Cerda, A., Artola, A., Font, X., Barrena, R., Gea, T., & Sánchez, A., (2018). Composting of food wastes: Status and challenges. *Bioresource Technology, 248*, 57–67.

Chau, C. F., Chen, C. H., & Lee, M. H., (2004). Comparison of the characteristics, functional properties, and *in vitro* hypoglycemic effects of various carrot insoluble fiber-rich fractions. *LWT-Food Science and Technology, 37*(2), 155–160.

Choi, S. H., Kozukue, N., Kim, H. J., & Friedman, M., (2016). Analysis of protein amino acids, non-protein amino acids and metabolites, dietary protein, glucose, fructose, sucrose, phenolic, and flavonoid content and antioxidative properties of potato tubers, peels, and cortexes (pulps). *Journal of Food Composition and Analysis, 50*, 77–87.

Crea, R., (2002). *Method of Obtaining a Hydroxytyrosol-Rich Composition from Vegetation Water.* World Intellectual Property Organization. WO/2002/0218310.

Crowely, P. H., & Brown, C. P., (2007). *Method of Deflavoring Whey Protein Using Membrane Electrodialysis.* European Patent Office. EP 1787528 (A1).

Davis, M. E., Nelson, L. A., Keenan, J. M., & Pins, J. J., (2003). *Reducing Cholesterol with Hydrolyzed Whey Protein.* World Intellectual Property Organization. WO/2003/063778.

De Moura, I. G., de Sá, A. V., Abreu, A. S. L. M., & Machado, A. V. A. (2017). Bioplastics from agro-wastes for food packaging applications. In *Food Packaging* (pp. 223–263). Academic Press.

De Moura, I. G., De Sá, A. V., Abreu, A. S. L. M., & Machado, A. V. A., (2017). Bioplastics from agro-wastes for food packaging applications. In: *Food Packaging* (pp. 223–263). Academic Press.

De Vuyst, L., De Vin, F., Vaningelgem, F., & Degeest, B., (2001). Recent developments in the biosynthesis and applications of heteropolysaccharides from lactic acid bacteria. *Int. Dairy J., 11*(9), 687–707.

Dhillon, G. S., Brar, S. K., Verma, M., & Tyagi, R. D., (2011). Enhanced solid-state citric acid bio-production using apple pomace waste through surface response methodology. *Journal of Applied Microbiology, 110*(4), 1045–1055.

Esparza, I., Jiménez-Moreno, N., Bimbela, F., Ancín-Azpilicueta, C., & Gandía, L. M., (2020). Fruit and vegetable waste management: Conventional and emerging approaches. *Journal of Environmental Management, 265*, 110510.

Ezejiofor, T. I. N., Enebaku, U. E., & Ogueke, C., (2014). Waste to wealth-value recovery from agro-food processing wastes using biotechnology: A review. *Biotechnology Journal International*, 418–481.

Farah, S., Anderson, D. G., & Langer, R., (2016). Physical and mechanical properties of PLA, and their functions in widespread applications — A comprehensive review. *Adv. Drug Deliv. Rev. 107*, 367–392.

Fernandez-Bolanos, J., Heredia, A., Rodr ~ ıguez, G., Rodrıguez, R., Guillen, R., & Jimenez, A., (2002). *Method for Obtaining Purified Hydroxytyrosol from Products and by-Products Derived from the Olive Tree.* World Intellectual Property Organization. WO/2002/ 064537.

Follonier, S., Goyder, M. S., Silvestri, A. C., Crelier, S., Kalman, F., Riesen, R., & Zinn, M., (2014). Fruit pomace and waste frying oil as sustainable resources for the bioproduction of medium-chain-length polyhydroxyalkanoates. *Int. J. Biol. Macromol., 71,* 42–52.

Friedman, M., (1997). Chemistry, biochemistry, and dietary role of potato polyphenols: A review. *Journal of Agricultural and Food Chemistry, 45*(5), 1523–1540.

Funami, T., Yotsuzuka, F., Yada, H., & Nakao, Y., (2006). Thermoreversible characteristics of curdlan gels in a model reduced-fat pork sausage. *J. Food Sci., 63*(4), 575–579.

Galanakis, C. M., (2012). Recovery of high added-value components from food wastes: Conventional, emerging technologies and commercialized applications. *Trends in Food Science & Technology, 26*(2), 68–87.

Ghosh, P. R., Fawcett, D., Sharma, S. B., & Poinern, G. E., (2017). Production of high-value nanoparticles via biogenic processes using aquacultural and horticultural food waste. *Materials, 10*(8), 852.

González-Centeno, M. R., Rosselló, C., Simal, S., Garau, M. C., López, F., & Femenia, A., (2010). Physico-chemical properties of cell wall materials obtained from 10 grape varieties and their byproducts: Grape pomaces and stems. *LWT-Food Science and Technology, 43*(10), 1580–1586.

Gorinstein, S., Martín-Belloso, O., Park, Y. S., Haruenkit, R., Lojek, A., Číž, M., & Trakhtenberg, S., (2001). Comparison of some biochemical characteristics of different citrus fruits. *Food Chemistry, 74*(3), 309–315.

Guangyu, H., & Xiaoyan, Z., (2011). *Method for Preparing Punicalagin and Ellagic Acid from Pomegranate Rind.* State Intellectual Property Office of the People's Republic of China. CN 101974043A.

Gupta, C., Prakash, D., Gupta, S., & Nazareno, M. A., (2019). Role of vermicomposting in agricultural waste management. In: *Sustainable Green Technologies for Environmental Management* (pp. 283–295). Springer, Singapore.

Gustavsson, J., Cederberg, C., Sonesson, U., Van, O. R., & Meybeck, A., (2011). *Global Food Losses and Food Waste.* Extent, causes and prevention. Rome.

Himaja, A. L., Karthik, P. S., Sreedhar, B., & Singh, S. P., (2014). Synthesis of carbon dots from kitchen waste: Conversion of waste to value-added product. *Journal of Fluorescence, 24*(6), 1767–1773.

Imandi, S. B., Bandaru, V. V. R., Somalanka, S. R., Bandaru, S. R., & Garapati, H. R., (2008). Application of statistical experimental designs for the optimization of medium constituents for the production of citric acid from pineapple waste. *Bioresour. Technol., 99,* 4445–4450.

Jensen, J., & Larsen, P. H., (1993). *Method for Obtaining High-Quality Protein Products from Whey.* World Intellectual Property Organization. WO/1993/021781.

Jishan, L., Jian, Z., Jianyong, G., Yang, L., Niannian, N., & Hongying, G., (2009). *Method for Extracting and Recycling Albumin from Whey Wastewater from Production of Soy Protein Isolate.* State Intellectual Property Office of the People's Republic of China. CN 101497645.

Kanimozhi, J., Sivasubramanian, V., Achary, A., Vasanthi, M., Vinson, S. P., & Sivashankar, R., (2018). Bioprocessing of agrofood industrial wastes for the production of bacterial exopolysaccharide. In: *Bioprocess Engineering for a Green Environment* (pp. 67–98). CRC Press.

Kapoor, M., Panwar, D., & Kaira, G. S., (2016). Bioprocesses for enzyme production using agro-industrial wastes: Technical challenges and commercialization potential. In: Dhillon, G. S., & Kaur, S., (eds.), *Agro-Industrial Wastes as Feedstock for Enzyme Production* (pp. 61–93). Academic Press – Elsevier, Amsterdam.

Kareem, S. O., & Rahman, R. A., (2013). Utilization of banana peels for citric acid production by *Aspergillus Niger. Agriculture and Biology Journal of North America, 4*(4), 384–387.

Kibler, K. M., Reinhart, D., Hawkins, C., Motlagh, A. M., & Wright, J., (2018). Food waste and the food-energy-water nexus: A review of food waste management alternatives. *Waste Manag., 74*, 52–62.

Kumar, H., Bhardwaj, K., Sharma, R., Nepovimova, E., Kuča, K., Dhanjal, D. S., & Kumar, D., (2020). Fruit and vegetable peels: Utilization of high value horticultural waste in novel industrial applications. *Molecules, 25*(12), 2812.

Kumar, K., Yadav, A. N., Kumar, V., Vyas, P., & Dhaliwal, H. S., (2017). Food waste: A potential bioresource for extraction of nutraceuticals and bioactive compounds. *Bioresources and Bioprocessing, 4*(1), 18.

Lavecchia, R., & Zuorro, A., (2008). *Process for the Extraction of Lycopene.* World Intellectual Property Organization. WO/2008/ 055894.

Manfredi, S., Tonini, D., & Christensen, T. H., (2010). Contribution of individual waste fractions to the environmental impacts from landfilling of municipal solid waste. *Waste Management, 30*(3), 433–440.

Mehta, K., & Duhan, J. S., (2014). Production of invertase from *Aspergillus Niger* using fruit peel waste as a substrate. *Intl. J. Pharm. Biol. Sci., 5*, 353–360.

Mrudula, S., & Anitharaj, R., (2011). Pectinase production in solid-state fermentation by *Aspergillus Niger* using orange peel as substrate. *Glob. J. Biotechnol. Biochem., 6*, 64–71.

Mudaliyar, P., Sharma, L., & Kulkarni, C., (2012). Food waste management-lactic acid production by *Lactobacillus* species. *Int. J. Adv. Biol. Res., 2*(1), 34–38.

Mueller, D., Jung, K., Winter, M., Rogoll, D., Melcher, R., Kulozik, U., & Richling, E., (2018). Encapsulation of anthocyanins from bilberries – effects on bioavailability and intestinal accessibility in humans. *Food Chemistry, 248*, 217–224.

Panda, S. K., Mishra, S. S., Kayitesi, E., & Ray, R. C., (2016). Microbial-processing of fruit and vegetable wastes for production of vital enzymes and organic acids: Biotechnology and scopes. *Environ. Res., 146*, 161–172.

Patra, J. K., Kwon, Y., & Baek, K. H., (2016). Green biosynthesis of gold nanoparticles by onion peel extract: Synthesis, characterization, and biological activities. *Adv. Powder Technol., 27*, 2204–2213.

Pirog, T. P., Ivakhniuk, M. O., & Voronenko, A. A., (2016). Exopolysaccharides synthesis on industrial waste. *Biotechnologia Acta, 9*(2).

Raji, Y. O., Jibril, M., Misau, I. M., & Danjuma, B. Y., (2012). Production of vinegar from pineapple peel. *International Journal of Advanced Scientific Research and Technology, 3*(2), 656–666.

Ravindran, R., & Jaiswal, A. K., (2016). Exploitation of food industry waste for high-value products. *Trends in Biotechnology, 34*(1), 58–69.

Reddy, C. S. K., Ghai, R., & Kalia, V., (2003). Polyhydroxyalkanoates: An overview. *Bioresource Technology, 87*(2), 137–146.

Robbins, R. J., (2003). Phenolic acids in foods: An overview of analytical methodology. *Journal of Agricultural and Food Chemistry, 51*(10), 2866–2887.

Safdar, M. N., Kausar, T., Jabbar, S., Mumtaz, A., Ahad, K., & Saddozai, A. A., (2017). Extraction and quantification of polyphenols from kinnow (*Citrus reticulate* L.) peel using ultrasound and maceration techniques. *Journal of Food and Drug Analysis, 25*(3), 488–500.

Said, A., Leila, A., Kaouther, D., & Sadia, B., (2014). Date wastes as substrate for the production of α-amylase and invertase. *Iran J. Biotechnol., 12*, 41–49.

Salemdeeb, R., Zu Ermgassen, E. K. H. J., Kim, M. H., Balmford, A., & Al-Tabbaa, A., (2017). Environmental and health impacts of using food waste as animal feed: A comparative analysis of food waste management options. *J. Clean. Prod., 140*, 871–880.

Sharma, R., Oberoi, H. A., & Dhillon, G. S., (2016). Fruit and vegetable processing waste: Renewable feedstocks for enzyme production. In: Dhillon, G. S., & Kaur, S., (eds.), *Agroindustrial Wastes as Feedstock for Enzyme Production* (pp. 23–59). Academic Press – Elsevier, Amsterdam.

Shenghui, Z., (1995). *Preparation of Chitosan Derivative Fruit and Vegetable Antistaling Agent*. State Intellectual Property Office of the People's Republic of China. CN 1106999.

Shilev, S., Naydenov, M., Vancheva, V., & Aladjadjiyan, A., (2007). Composting of food and agricultural wastes. In: *Utilization of By-products and Treatment of Waste in the Food Industry* (pp. 283–301). Springer, Boston, MA.

Skiba, M. I., & Vorobyova, V. I., (2019). Synthesis of silver nanoparticles using orange peel extract prepared by plasmochemical extraction method and degradation of methylene blue under solar irradiation. *Adv. Mater. Sci. Eng., 2019*, 1–8.

Soong, Y. Y., & Barlow, P. J., (2004). Antioxidant activity and phenolic content of selected fruit seeds. *Food Chemistry, 88*(3), 411–417.

Steinbüchel, A., (2001). Perspectives for biotechnological production and utilization of biopolymers: Metabolic engineering of polyhydroxyalkanoate biosynthesis pathways as a successful example. *Macromol. Biosci., 1*(1), 1–24.

Tornberg, E., & Galanakis, C. M., (2008). *Olive Waste Recovery*. World Intellectual Property Organization. WO/2008/082343.

Tsang, Y. F., Kumar, V., Samadar, P., Yang, Y., Lee, J., Ok, Y. S., & Jeon, Y. J., (2019). Production of bioplastic through food waste valorization. *Environment International, 127*, 625–644.

United Nation, (2015). *Transforming Our World: The 2030 Agenda for Sustainable Development*. [WWW Document]. http://www.un.org/ga/search/view_doc. asp?symbol=A/RES/70/1&Lang=E (accessed on 30 December 2021).

Vijayakumar, J., Aravindan, R., & Viruthagiri, T., (2008). Recent trends in the production, purification, and application of lactic acid. *Chem. Biochem. Eng., 22*(2), 245–264.

Vikas, O. V., & Mridul, U., (2014). Bioconversion Of papaya peel waste in to vinegar using acetobacter aceti. *International Journal of Scientific Research, 3*(11), 409–411.

Vikneswaran, R., Ramesh, S., & Yahya, R., (2014). Green synthesized carbon nanodots as a fluorescent probe for selective and sensitive detection of iron (III) ions. *Materials Letters*, 179–182.

Wadhwa, M., & Bakshi, M. P. S., (2013). *Utilization of Fruit and Vegetable Wastes as Livestock Feed and as Substrates for Generation of Other Value-Added Products* (Vol. 4, pp. 1–67). Rap Publication.

Zhang, H., Chen, J., Li, J., Wei, C., Ye, X., Shi, J., & Chen, S., (2018). Pectin from citrus canning wastewater as potential fat replacer in ice cream. *Molecules, 23*(4), 925.

Zhang, H., Nishinari, K., Williams, M. A. K., Foster, T. J., & Norton, I. T., (2002). A molecular description of the gelation mechanism of curdlan. *Int. J. Biol. Macromol., 30*(1), 7–16.

Zhao, J., Huang, M., Zhang, L., Zou, M., Chen, D., Huang, Y., & Zhao, S., (2017). Unique approach to develop carbon dot-based nanohybrid near-infrared ratiometric fluorescent sensor for the detection of mercury ions (Article). *Analytical Chemistry*, (No. 15), 8044–8049.

CHAPTER 4

Basic and Modern Environmental Management Practices for Food and Agricultural Waste Management

QURAAZAH AKEEMU AMIN, TOWSEEF AHMAD WANI,
TAWHEED AMIN, AFSAH IQBAL NAHVI, TAHA MUKHTAR,
NAZRANA RAFIQUE, and SHUBLI BASHIR

Division of Food Science and Technology, Sher-e-Kashmir University of Agricultural Sciences and Technology of Kashmir, Jammu and Kashmir – 190025, India, E-mail: widaad57@gmail.com (Q. A. Amin)

ABSTRACT

In today's world, food wastage has become a crisis. A huge amount of waste is generated on a daily basis which is a serious concern affecting both developed as well as developing countries equally. In developed countries, the main contributors to food waste are consumers because of incorrect food-management habits and behavior, and ultimately, they have to reimburse wastes that are identified throughout the food chain. FAO reports the tierce of the total amount of produce that is being augmented for human utilization remains unconsumed and hence, therefore, is waste. Agricultural wastes are the leftover or non-product outputs having low economic values that are obtained from the production and processing of agricultural products having a huge impact on the profitability of the whole food supply chain (FSC). These wastes are usually generated from

---

Integrated Waste Management Approaches for Food and Agricultural Byproducts. Tawheed Amin, PhD,
Omar Bashir, Shakeel Ahmad Bhat & Muneeb Ahmad Malik (Eds.)
© 2023 Apple Academic Press, Inc. Co-published with CRC Press (Taylor & Francis)

cultivation, aquaculture, and livestock. Thus the excessive proportion of food waste is being generated by the foodservice sector manifests substantial environmental and financial consequences. Currently, "3R" strategy is used for optimization of waste from production and processing sector.

4.1 INTRODUCTION

Food is indispensable and its loss is one of the great challenges faced by humanity. In this day and age hunger is pervasive. However, more than 20,000 people die of hunger and one among nine sleep hungry (Prabhash, 2017). Food loss (FL) in the industry is a vital problem. It is strenuous to change the operation that leads to food wastage. Globally 70% of such waste is dumped into landfill sites that lead to evolution of methane which promotes global warming. In India, 40% of the food produced gets dissipated and the wastage accounts for about $14 billion a year (Uzmi, 2018). This not only poses economic threats but manifests a grave environmental repercussion. As we know, there are limited natural resources, so modest solutions are a prerequisite from sourcing of raw materials to the distribution to consumers in the management practices of the supply chain. Food waste is defined as the *"loss of materials planned for human ingestion that are afterward either discharged, which thereby get contaminated, degrade, and are subsequently lost"* (Anupam et al., 2019). However, it includes a description of FSC production stage along with postharvest and processing stage (FAO, 2011). Food waste is also defined as, *"the loss of food taking place either at the market stage or at final consumption and utilization stages and is generated due to the negligent behavior on the part of retailer as well as consumer"* (Parfitt et al., 2010). European Project FUSIONS defines food waste as *"any form of food, edible or inedible, aloof from (diverted or lost from) the FSC that is to be either disposed or improved (includes AD, incineration, composted crops, co-generation, bio-energy production, sewer disposal, landfill or discarded into the sea)"* (Ostergren et al., 2014). Agricultural waste is defined as, *"the residue arising from the production and processing of raw products such as fruits, vegetables, crops, poultry, dairy, etc."* These are the outputs of leftover or non-products having low economic values that are obtained from the formation and processing of agricultural products with a huge impact on the profitability of the whole supply chain for food. The composition depends upon the system and agricultural activities and it is in the form of slurries, liquid, and solids. Agricultural waste also known as agro-waste comprises of food processing waste (20% of maize canned and 80% waste), animal waste (carcass and manure), crop waste (sugarcane bagasse, corn stalks, and culls from fruits and vegetables) and toxic agricultural wastes (insecticides, pesticides).

Distinguishing between edible with non-edible food parts is always subjective and not straightforward. The portion which is edible for some parts of the world may be inedible for other parts (Redlingshofer and Soyeux, 2012; Buzby and Hyman, 2012). In most cases, the dissimilarity between edible and non-edible depends solely on the dietary habits (consumption of apple peel, bread crust, fat on meat, etc.), food culture and geographic location. So the food that is not consumed like inedible part is considered to be *"food waste."*

The research carried out in the field of agriculture demonstrates that the loss in foods is attributed to climatic and environmental dynamics, diseases, and parasites. But there are differences when we compare developed and developing countries in terms of infrastructures, sowing, cultivation, harvesting, storage, and processing. Economic factor plays an important role in both developed as well as in developing countries. The primary reason of food waste during processing is incompetence and technical breakdown in production processes known as *"production waste."*

In commercial countries, the main reason contributing to waste occurs in the final phase of FSC, such as restaurants, domestic consumption and foodservice organizations. However, FL documented at agricultural levels in such countries is not insignificant, for instance. In Italy (2009), 17.7 million tons of agricultural foodstuff was left within fields constituting 3.25% of total production (Segre and Gaiani, 2011).

The waste obtained from agriculture and agro-industrial sources comprises of bulky solids, airborne contaminants, and wastewater. Inappropriate decomposition of it results in contamination of land, air, and water. Wastewater is of most concern as food processing involves unit operation like washing, evaporation, filtration, and extraction. These contain higher concentrations of soluble organics (carbohydrates, lipids, and proteins) and suspended solids that are difficult to dispose off. Table 4.1 shows characteristic of wastewater from food industries.

116   Integrated Waste Management Approaches for Food and Agricultural Byproducts

TABLE 4.1   Comparative Strengths of Wastewaters from Food Processing Industries

By-Products	BOD (mg/L)	TSS (mg/L)	pH	Protein (mg/L)	Fat (mg/L)
Dairy	1,000–4,000	1,000–2,000	Acid	6–82	30–100
Farm	1,000–2,000	1,500–3,000	75.5–8.5	–	–
Fish	500–2,500	100–1,800	–	300–1,800	100–800
Fruit	1,200–4,200	2,500–6,700	Acid	–	–
Meat	1,000–6,500	100–1,500	–	350–950	15–600
Poultry	500–800	450–800	6.5–9.0	300–650	100–400
Silage	50,000	Low	Acid	–	–
Slaughterhouse	1,500–2,500	800	11–13	–	–
Vegetable	1,000–6,800	100–4,000	Acid	–	–

Abbreviations: BOD: Biochemical oxygen demand; TSS: Total soluble solids.

Source: Conly et al. (2019).

4.2   AGRICULTURAL WASTE MANAGEMENT SYSTEM (AWMS)

The agricultural waste management system (AWMS) is a planned system in which all necessary components are installed and managed in order to control and use agricultural production by-products to maintain and enhance the quality of resources such as air, water, soil, and plants and animals resources (USDA, 2013). This type of system uses a total system approach that is intended to be used throughout the year for all types of wastes associated with agricultural production. The concentration of agricultural waste in total solids determines the handling of material and affects the total volume of waste. Environment, animal type, amount of water consumed by the animal and the type of feed are the variables that influence the total solids concentration. The total solids concentration increases by adding beddings or other solid wastes, decreases by the addition of water and stabilizes by giving protection from additional water (Zhongzhong et al., 2020) (Figure 4.1).

4.3   PRINCIPLE OF WASTE MANAGEMENT

Human day-to-day activities are responsible for the introduction of waste into the environment. WM is the practice, method, or process of handling waste at every phase including production, capturing, and discarding.

Basic and Modern Environmental Management Practices for Food

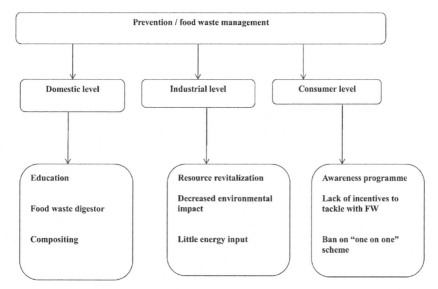

FIGURE 4.1 Flowchart for food waste management.

To prevent the contact of waste with community and environment it is must to manage the waste thereby, safeguarding both. Depending on the physical state of waste, it is categorized as liquid or solid, hazardous or non-hazardous. Liquid waste consists of human waste (feces and urine), industrial waste sullage and run-off (gray water and rainwater). Solid waste includes residential waste, refuse, household waste, institutional (offices, schools, universities), and agricultural waste. **Hazardous waste is** the waste, *"that is classified on the basis of their properties and potential to cause harm. They may be corrosive, toxic, reactive, ignitable, and infectious."* The potential source of hazardous waste in rural household includes obsolete herbicide, pesticides, and rodenticides.

4.4 FUNCTIONS OF AGRICULTURAL WASTE MANAGEMENT SYSTEM (AWMS)

1. **Production:** The amount and nature of agricultural waste generated needs to be managed if the amount is too high. Production analysis involves consistency, type, quantity, timing as well as location of waste generated.

2. **Collection:** This is the collection and assembly of waste resulting from the deposition site. AWMS plan identifies the method of capturing, location of capturing waste it's scheduling, labor requirements, necessary facilities, management, and installation costs of components and finally the impact on the consistency of waste that is collected.
3. **Storage:** It deals with the containment and holding of waste temporarily. The control that is provided by storage over the arrangement and timing of device operations include handling and application of waste generated that could be affected by weather or by other processes (Figure 4.2).

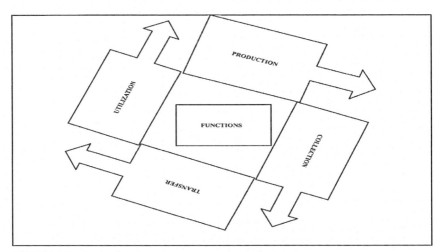

FIGURE 4.2 Agricultural waste management functions.

4. **Treatment:** The main function of treatment is that it reduces the contamination or toxicity of waste that includes physical, chemical, and biological treatment and moreover increases its potential beneficial use. Here pre-treatment activities that include analysis of the characteristics of waste prior to treatment; a determination of the desired characteristics of waste after treatment; the selection of the type, estimated size, location, and the installation cost of the treatment facility; and the management cost of the treatment process are embraced.

5. **Transport:** Depending on TS concentration in solids, liquids, and slurries, the waste is transferred from collection point to the operation point.
6. **Operation:** The waste is recycled and reintroduced into the environment for useful purposes (Purabi et al., 2016).

4.5 IMPORTANCE OF WASTE MANAGEMENT

The waste generated should be treated with care as it can be harmful to humans and can cause various diseases. So it becomes the responsibility of society to ensure waste disposal in its best practice, and the initiative is taken by collecting different types of wastes in separate bins. The significant role is played by the government in ensuring sensitization of people as well as creating policies for guiding people about waste disposal practices and also ensures its implementation (Figure 4.3).

According to the Indian Constitution, WM falls within the purview of state list as it is a part of public health and sanitation. It has been estimated that 38 million tons of solid waste is generated by 300 million people of urban India, according to estimates. Therefore, it becomes necessary to dispose of waste properly so as to improve and ensure public health (Jaspal et al., 2018).

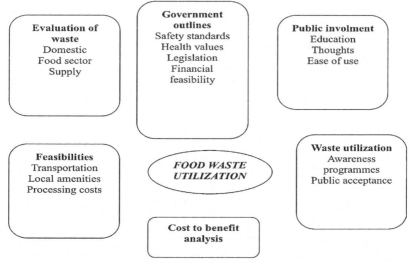

FIGURE 4.3 Factors to be considered for utilization of food waste.

In urban India, Municipal corporations are the local bodies accountable for managing waste and related activities associated to the wellbeing of the community. Nowadays, due to political and public awareness the WM system has started receiving attention and NGO's, have also taken initiative from the past few years (Singh et al., 2018).

4.6 CONVENTIONAL FOOD WASTE MANAGEMENT TECHNOLOGIES

4.6.1 ANIMAL FEEDING

Direct or pre-processed FL and waste can be used as animal feed such as silage, fluid feeding, etc. This choice ought to be considered cautiously as every animal cannot consume every food/residue. In addition, the danger of poisonous mixes causing illness is an adverse outcome. Intermittently it does not suffice the dietary perspective. In addition, the transportation, and at last preservation costs related make this option much of the time unworkable. Thermally pre-treated food waste is an adequate alternative but it raises the expenses. Animal feeding is obviously favorably contrasted with different choices, for example, landfilling. Advantages principally come from the partial replacement of typical feed, the creation of which has critical ecological and health effects (UgoDe Corato, 2020).

4.6.2 LANDFILLING

It is the most widely recognized and straightforward approach to discard solid deposits despite its ecological effect. This is mainly due to food waste high biodegradability, which is associated with a high nutrient content. Methane evolving from landfills caused by the natural anaerobic eruption of the unload is the third largest anthropogenic source of discharge of atmospheric methane. The produced gas, otherwise referred to as biogas (blend of CH_4 and CO_2), is regularly radiated to climate. Notwithstanding, an ever increasing number of locales are catching biogas because of lawful prerequisites and its potential as fuel to create electricity or cogenerate heat and force in motor-generators. Moreover biogas can perhaps be additionally processed to isolate CO_2 and other minor segments to acquire bio-methane.

Basic and Modern Environmental Management Practices for Food   121

4.6.3 ANAEROBIC DIGESTION

Anaerobic digestion (AD) is a demonstrated innovation towards change of waste materials into functional gas and power. However, it is also a notable innovation for the administration of excrement and of slime created in water resource recovery facility (WRRF). Normal processes involved in AD of organic matter comprising food waste are virtually similar to those that occur in landfill. The primary distinction is that the AD is completed in explicitly planned bioreactors that permit controlling the fundamental working boundaries. Especially the control of temperature is significant as the anaerobic microorganisms included are mesophiles or thermophilic, being active within temperature range of 30–60°C, respectively. Biological degradation is perplexing cycle that included few stages as given in subsections.

4.6.3.1 HYDROLYSIS

During this progression, hydrolysis of macromolecules, for example, polysaccharides, proteins, and nucleic acids, occurs by fermentative microorganisms creating organic substances like amino acids, sugars, and fatty acids.

4.6.3.2 ACIDOGENESIS

Throughout the subsequent advance, acid-forming microorganisms transform the hydrolytic items into volatile fatty acids (VFA's), predominantly acetic, propionic, butyric, and valeric acids, alcohols, ammonia, CO_2 and hydrogen sulfide.

4.6.3.3 ACETOGENESIS

The subsequent stage includes change by syntrophic acetogens of the higher organic acids and alcohols into acetic acid, carbon dioxide and hydrogen.

4.6.3.4 METHANOGENESIS

In this final step, methane is formed by the activity of two groups of methane producing bacteria; either of them parts acetic acid derivation particles into methane and CO_2, which further produces methane from the transformation of CO_2 and hydrogen. The worldwide cycle is constrained by the slow pace of the hydrolysis step, as hydrolytic catalyst should get adsorbed on the surface of the solid organic substrates. Therefore, it is gainful to pre-treat the substrates, for instance, by physical methods (mechanical grinding, microwave or ultrasounds treatments), with the point of segregating particles and lessening their size. Chemical, thermal, and biological pre-treatments have also been useful. In single-stage digesters run in semi-continuous mode, AD is usually performed in which all responses occur at the same time with few stages involving the break-down of organic matter. Methane found in biogas is the most essential product of AD, but the digestate obtained as a by-product of the digestion process may also be valorized, either directly as animal bedding and as a substrate in soil modifications, or by subsequent treatments of composting or vermicomposting to obtain manures.

4.7 COMPOSTING AND VERMI-COMPOSTING

Soil fertilization is an important strategy for waste bioconversion, with vapor discharges and impurities being its key obstacles. The nitrous oxide and methane emitted during soil fertilization contribute enormously to global warming. Their influence is estimated to be 310 and 20 times higher than CO_2, respectively. Food waste accounts for almost half of the total municipal waste in most countries, though this proportion may be higher in developing countries. Fruits, vegetables, processed food waste; beef, etc. are organic components of food waste. During cultivation, handling, storage, processing, and consumption, food waste is produced. FL and waste contain a huge amount of organic matter and moisture thus decays effectively. Thus, composting can be considered to be an environmentally friendly and viable management option for FL and waste. It is an innovation that takes into account the administration of wastes organic matter, and consequently it is appropriate for food waste and especially for fruit and vegetable waste (FVW). Environmental benefits result from lower emissions of greenhouse emission and leachate production compared to landfilling. Although, few imperatives are required: sorting of the wastes to isolate the natural portion, energy for blending and air circulation, water incorporation and a cautious control of the working conditions.

Composting requires the aerobic biological transformation of organic material, releasing CO_2, nitrogen, and ammonia and a solid recalcitrant material also referred to as compost containing humic substances compared to the processes found in landfills. Compost is used as a kind of improvement in the soil that helps to preserve water, thereby reducing irrigation requirements. This use, however, includes the absence of heavy metals, pathogens, and micropollutants. Potential of hydrogen, C/N ratio, moisture content, air circulation rate, molecule size, and porosity are basic composting process variables (level of compaction of the organic matter). Odor emissions can lead to important environmental effects under inappropriate conditions and, at the same time, a fertilizer of inferior quality is acquired. For the development of bio-manures, vermicomposting of food waste has been introduced, which involves the adaptation of waste organic matter through the cooperation of earthworms and microorganisms. In terms of nutrient accessibility in the soil, the acquired end product (vermicompost) has favored properties over traditional compost. In some recent research, the earthworm activity of vermicompost from food waste has been studied. It can be argued that vermin-composting can successfully kill the bacterial pathogenic burden, making it more suitable for soil manure production than composting (Chowdhury et al., 2014).

4.8 THERMAL TREATMENTS

Incineration, or combustion, is an operation that is favorable from the perspective of critically lowering in waste volume finally discarded. Thus, it helps in expanding landfills' lifetime. The incineration plant ought to be correctly planned and operated to protect pollution of air, however, modern incineration plants coordinate reasonable air contamination control innovations bringing about incredibly low outcome of toxins. Because of a large amount of energy required, its installation and operation costs are high. This is due to low calorific substances and high food waste moisture content, and in particular FVW, which imposes the necessity for food waste to be co-terminated with other high-energy waste. In addition to the risk of developing very harmful pollutants (including dioxins), net CO_2 discharges into the climate increase dramatically when plastics are used. Combustion heat may be partially recovered and used, depending on the requirements and interests, in the incineration phase itself (for example, to preheat air) or for power generation. Despite the fact that harmful species of nitrogen oxide are generated with the flue gases, peels obtained from orange can be effectively valorized by combustion. FL and waste can be easily pelletized for use in a regular residential stove with temperatures and emissions comparable to wood and can be efficiently burned. In any event, attention should be given to the capital expenditure and maintenance of equipment for processing (grinding), drying, and properly and safely burning or gasifying waste food. Despite the fact that applications have been developed to use ashes as building material, an inorganic fraction of the solid fuels used, which is most commonly landfilled.

Incineration is the primary innovation for the management of solid waste (a share of approximately 80%), due to the pressure posed by the scarce accessibility of land, which prompts landfilling restriction strategies. Gasification and pyrolysis are various forms of thermal treatments in which the organic matter is converted into various materials. In both procedures, among other elements, the vaporous stream produced contains hydrogen and CO_2, which can be additionally processed to obtain blend gas, otherwise referred to as syngas (a mixture of hydrogen and carbon monoxide in different areas) that can be used as feedstock for the synthesis of chemicals, or can be used directly as fuel for the generation of power and/or heat. On account of pyrolysis, in a non-oxidizing environment and without any other reactant, the thermal decomposition of the waste is oriented towards the processing of three fractions (gas, liquid, and solid), the degree of which depends on working conditions, i.e., the process temperature, the heating rate and the vapor residence time. There are a few uses and applications for solid (char) and liquid (pyrolysis oil or bio-oil) fractions, including soil alteration, for acquiring chemicals and for producing clean syngas. The method is incomplete oxidative as regards gasification, providing a gas stream (i.e., producer gas) that can be used as flammable or further refined into syngas. As for food waste and food and vegetable WM, these developments are only at an early stage of improvement (Esparza et al., 2020).

4.9 THE 3R'S STRATEGY

The 3R's strategy can be imperative and reasonable alternative for diminishing waste at a least level. "*3R Initiative*" was reliably sent to 3R Conference of Ministers facilitated by the Japanese Authority in April 2005, with

Basic and Modern Environmental Management Practices for Food 125

the aim to bring growth in the activity on 3R globally. A high level meeting on 3R was coordinated in Japan in March 2006, generating strong commitment of the state and different stakeholders to implement 3R at local, federal, and regional stages. The waste minimization can be beneficial if the 3Rs are suggested in the hierarchical order which is 'Reduce,' 'Reuse,' and 'Recycle' (Figure 4.4). 3R strategy is traditionally passed on with the assistance of a pyramid hierarchy where the development of ecological preferences of every strategy is suited from the base to the top (Plazzotta et al., 2017).

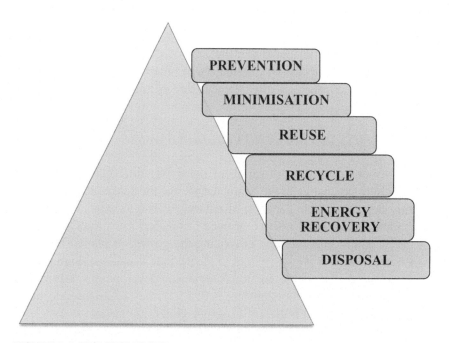

FIGURE 4.4. 3R'S HIERARCHY.

4.9.1 ADVANTAGES OF 3R'S

The 3Rs have a positive impact on the community and climate:

i. Greenhouse gases (GHG) decreased by energy effectiveness and asset proficiency and may diminish the CO_2 discharge.
ii. Method transformation, employment, and opportunity creation.
iii. Natural gas is produced from bio-wastes.

iv. Prevent contamination and improves edaphic parameters and may give a solid climate for individuals of the urban areas.
v. Financially it can be beneficial for creating recyclable force and power.

4.10 REDUCTION OF FRUIT AND VEGETABLE WASTE

This is of main concern in the waste order that generally rely on practice of development. Any of these can't be effectively altered. For instance, crop production has to be fundamentally greater than trade prediction, to confront possible reaping deprivation due to usual occurrence. Large amount of fruit and vegetables are wasted of the fact that items do not satisfy standard norms set by distributors or buyers. These small-sized or deformed products are typically characterized as "substandard."

4.10.1 REUSE OF FRUIT AND VEGETABLE WASTE

Reuse demonstrates the utilization of debris for different causes either with slight or no change in the characteristics of the debris. Reusing FVW directly depends on the capacity of natural residue to expand the nature of contaminated soil by restricting trace metals and metalloids, impeding exchange to soil moisture and advancing the growth of plants. Nonetheless, reuse system is regularly hard to incorporate because of high natural insecurity of FVW, liable for microorganism development hazard and off-odors generation. Fiber substances of FVW may be misused to make animal fodder with high dietary benefits. This reuse system is restricted by certain limitations. The large amount of water content, frequently surpassing 80% leaving this debris is inclined to microbial pollution thus, removal of some part of moisture is typically needed. Moreover, substances with less protein content and more availability of inedible mixes are not generally reasonable for animal fodder (Salihoglu et al., 2018).

4.10.2 RECYCLE OF FRUIT AND VEGETABLE WASTE

The recuperation of debris after a significant adjustment in the attributes of the same is characterized as recycle. Recycle of FVWs offer in this manner a bigger number of potential outcomes than being reused. Recycle systems for wastes may be partitioned into methodologies where entire debris mass is recycled (composting, handling to flour, transformation into water) and procedures where explicit mixes are removed. Aerobic composting is an old eco-accommodating strategy to change over natural residues into natural compost. Nonetheless, it is grounded that anaerobic processing is a much appealing system to deliver compost from FVW, because of production of natural gas. The high contact surface and fibred structure of FVW flour had been utilized to imbibe toxins for example dyes and heavy metals from water and ground. To this respect, imbibitions are because of physical trapping into the permeable vegetable structure and to explicit communication with the functional groups of cellulose, hemi-cellulose, and lignin. FVW flour has likewise been utilized as an element for making items rich in functional compounds such as polyphenols and fiber. The fundamental preferred position of this recycle system is that important items, for example, adsorbents, and functional flours are acquired from minimal effort crude substances. No debris has to be discarded when prepared to flour. Nonetheless, the principle problem is significant expenses needed for removal of moisture from FVW due to the presence of the huge amount of moisture is present (Sindhu et al., 2019).

4.11 CONVENTIONAL WASTE MANAGEMENT PRACTICES

Conventional technologies basically follow a *"finish-of-pipe"* application, implying that their point is to limit the effect and harm caused to humans as well as climate, thus lastly discarding the debris previously produced. Given the buyers' bringing issues to the light on these issues, the food creation area should be especially delicate toward any activity prompting improved sustainability and decreased negative natural and biological effects of their exercises. Along these lines, 'green' or supportable creation of food ought to be considered of key significance for the future advancement of this key financial area.

4.12 FOOD WASTE AND MANAGEMENT

Food waste is a practice of not properly consuming the food. According to the food initiative taken by the United Nations (UN), food waste reflects a substantial decline in the quality and quantity of food. This means not only wasting the food to be eaten, but also wasting tremendous quantities of energy and time spent on its creation (Anupam et al., 2019). It has a significant contribution to greenhouse gas emissions and pollutes the water and other resources it takes to grow food. During the steps involved in the food systems, which include cultivation, processing, distribution, retail, and consumption, there are numerous reasons for FL. It plays a crucial role and has a direct impact on climate change and agriculture. The food waste is witnessing unprecedented growth and has therefore contributed to a global food crisis.

Food WM is a technique, methodology, or practice of utilizing the food properly and not wasting the food and the time, energy spend on it. Food WM is an approach which lays emphasis on how food can be saved from getting wasted. It is a way of reducing food waste for the environment and the individual. With the rise in population, huge tons of food are getting wasted as they don't reach to the human mouths and thus food systems are becoming inefficient. There is a dire need to switch to some methodology (food waste management) which will minimize the food waste and in turn can save the amount of energy and time spent on it. The food WM comprises of both conventional methodology and novel methodology (Anupam et al., 2019):

1. **Reduce:** The food taken by an individual should be minimum and hence should be consumed easily. The food should be cooked as per the requirement.

2. **Reuse:** The food to be cooked should be taken twice or thrice a day. It should also be used to make some other dishes.

3. **Recycle:** The food leftover after consumption should be recycled in a way that it should produce the products beneficial for commercial use.

4.13 NOVEL TECHNIQUE IN FOOD WASTE MANAGEMENT

4.13.1 ANAEROBIC DIGESTION

It is a process in which bacteria break down the food waste and produce biogas. This gas is used for biofertilizer and electricity. Anaerobic digester is used to carry out this process in the absence of oxygen. This process also results in the generation of methane and CO_2. The generated methane through this process is a proper solution to food WM. This process is cheap and produces less waste as residue (Nasir et al., 2012; Morita and Sasaki, 2012). This process is carried out through three steps and is depicted in Figure 4.5.

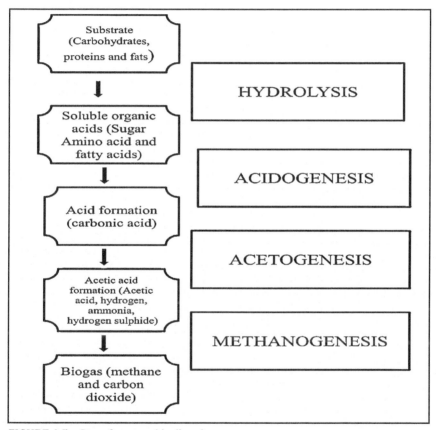

FIGURE 4.5 Steps for anaerobic digestion.

4.13.1.1 ENZYMATIC HYDROLYSIS

The polymers are broken down into smaller monomers or monomeric units in this process. The breakdown of polymers into monomers and dissolving them into a solution is known as hydrolysis. The complex organic molecules can be broken into fatty acids, sugars, and amino acids through hydrolysis (Li et al., 2011).

$$nC_6H_{10}O_5 + nH_2O \rightarrow nC_6H_{12}O_6 \qquad (1)$$

$$\text{Cellulose} \quad \text{Water} \quad \text{Glucose}$$

The optimum range of pH and temperature lies between 5–7 and 30–50°C, respectively.

4.13.1.2 ACIDOGENESIS

In this step, monomers are converted into fatty acids. The acidogenic microorganisms absorb by-products of hydrolysis through membrane (Li et al., 2011). The amount of fatty acids depends on digester. In this step, fermentation of the by-products of hydrolysis takes place (Kunwar et al., 2017).

4.13.1.3 ACETOGENESIS

In this step, organic acids and CO_2 are reduced to produce acetate. The bacteria which carry out this step are known as acetogenins. During acetate reduction, the percentage of methane rises to 70. Moreover, the percentage of hydrogen formed is 11.

$$nC_6H_{12}O_6 \rightarrow 3n\ CH_3COOH \qquad (2)$$

$$\text{Glucose} \quad \text{Acetic acid}$$

4.13.1.4 METHANOGENESIS

It is the last step of the AD in which methanogens carry out the process of methanogenesis, which belongs to Archaea. Reduction in CO_2 or acetic acid fermentation can result in the formation of methane. The by-products of step 1 and step 2 which act as precursor for the production of methane. The percentage of methane produced methanogens with the aid of carbon dioxide is 30.

$$CH_3COOH \rightarrow CH_4 + CO_2 \qquad (3)$$

$$\text{Acetic acid} \quad \text{Methane} \quad \text{Carbon dioxide}$$

Methanogens of two types produce methane in two ways, including acetoclastic (methanogens which uses acetic acid to form methane) and hydrogenotrophic (methanogens reduce carbon dioxide by utilizing hydrogen):

$$CO_2 + 4H_2 \rightarrow CH_4 + 3H_2O \qquad (4)$$
$$\text{Carbon dioxide} \quad \text{Hydrogen} \quad \text{Methane} \quad \text{Water}$$

4.13.2 AEROBIC DIGESTION

The process is carried out by the microorganisms with the aid of oxygen which oxidizes and CO_2 decomposes the organic components of stream of waste. In this technique, organic matter is rapidly consumed and converted into CO_2 by bacteria. This process is used to remove pathogenic microorganisms which can either be dry or wet. Wet aerobic digestion is similar to composting. Wet aerobic digestion is a fast-growing technology which is being carried out in the following steps as pulping, heating, aeration, inoculation with microorganisms and separation of products of solid, liquid fertilizer.

4.13.3 LIQUEFACTION

It is the process of formation of liquid from solid waste. It can be employed by different methods which include hydrothermal, enzymatic, and biological.

4.13.3.1 HYDROTHERMAL LIQUEFACTION (HTL)

It is a thermal depolymerization process which converts wet mass into oil with the aid of mild temperature and high pressure. The oil produced by this process has a high energy density with low heating value. Homogeneous/heterogeneous catalyst is used to improve the quality of the final product. The biomass of organic material containing carbon and hydrogen are thermochemically converted into hydrophobic compounds with high solubility and low viscosity.

4.13.3.2 BIOLOGICAL LIQUEFACTION

This process uses mechanical grinders to fragment food waste, which in turn is blended with water and additives for decomposition. These are mostly used for food wastes. It involves a blend of air, water, and micro-organism which speeds up the decomposition of waste in turn getting converted into liquid.

4.13.3.3 ENZYMATIC LIQUEFACTION

In this process, an enzyme, namely alpha-amylases, breaks down the wide chain of starch molecules into smaller ones which in turn results in the saccharification later producing the final product.

4.13.3.4 HYDROLYSIS

It is the conversion process in which simple sugars (glucose) are formed from cellulose which is present in organic matter. The sugars are then further fermented to ethyl alcohol. Cellulose can be splitted by various means which includes enzymes or acids.

4.13.3.4.1 Acid Hydrolysis

In this technique, concentrated or dilute acid is used for splitting of cellulose. Biomass is first crushed and then treated in a dilute acidic medium with process temperature around 450°F. Sulfuric acid in its concentrated form may also be used at the initial stage to de-crystallize cellulose before onset of dilute acid process.

4.13.3.4.2 Enzymatic Hydrolysis

The process involves the use of enzymes which are obtained from common fungi. These process have not been proven commercially yet but when they will be proven they will have significant cost advantage over the acid hydrolysis.

Basic and Modern Environmental Management Practices for Food 133

4.13.3.5 GASIFICATION

The biomass conversion process is called gasification (H_2) into carbon monoxide (CO), carbon dioxide (CO_2) and hydrogen (H_2). It is obtained when a material above 700°C temperature reacts with the steam or oxygen. Heating process can either be direct or indirect depending upon the configuration of system. Technologies for gasification have been widely demonstrated and are now used on a commercial scale. It has been mostly used commercially for processing of poultry wastes.

It is possible to further classify gasification into Plasma Assisted Gasification (PSG). Plasma is one of the states of matter and exists in the form of ionized gas. It is formed with the aid of plasma torch. An extensive amount of energy is generated with the aid of plasma which is not generally possible by fission/fusion (Amin et al., 2012). Earlier it was used for disintegration of wastes, but now is being used for solid waste treatment. This process is cheap and is used to produce an extensive amount of energy as an outcome. In this process evaporation of solid wastes takes place in the absence of oxygen. The waste is degraded and oxidized to carbon monoxide (CO), hydrogen (H_2) and water (H_2O). Organic matter waste is converted into synthetic gas, known as SynGas, and the left inorganic portion is transformed without ash into inert vitrified glass.

There are various techniques of PSG such as, SynGas polishers, waste zappers, and plasma-assisted gasification.

4.13.3.6 PYROLYSIS

It is a process of decomposing waste materials with the aid of heat at higher temperature (> 400°F) in the absence of oxygen. Greek terms 'pyro' means Fire and 'lysis' means splitting or disintegrating into smaller particles. It is a wood charring technique. It is of three types:

1. **Fast Pyrolysis:** In this process, the residues of biomass are heated at a high temperature with high rate of heating in absence of oxygen. They produce 60–70% liquid biofuels and 15–25% biochar depending on the biomass initial weight. They also yield 10–20% gaseous fuel which depends on the rate of biomass used.

2. **Flash Pyrolysis:** This process converts biomass into solid, liquid, and gaseous products. It is used to produce bio-oil. In this process,

devolatilization occurs in an inert atmosphere with higher heating rate and a temperature range of about 450 and 1,000°C. It has a poor thermal stability.

3. **Slow Pyrolysis:** This system produces high quality charcoal with low temperatures and low heating rates. This process produces low quality bio-oil.

$$C_6H_{10}O_5 \rightarrow 3.74C + 2.65H_2O + 1.17CO_2 + 1.08CH_4 \qquad (5)$$

Cellulose   Carbon   Water   Carbon dioxide   Methane

4.13.3.7 RENDERING

It is a process where conversion of waste animal tissue occurs into stable and useable forms. It can be referred to processing of products of animals in more useful items or to provide the whole lipid tissue in clarified lipids like lard or tallow. This process involves drying of material and separation of lipid from bone and protein. It results in the formation of lipid commodity and protein meal. Food waste, consisting of animal by-products such as fat oils, is contained in packaged products such as high-protein meat, tallow or grease, which are further used for soap, cosmetics, paint, toothpaste, lubricants, biofuels, etc. This process is classified into two types: edible rendering process and inedible rendering process.

4.13.3.7.1 Edible Rendering Process

It involves formation of edible fats and proteins in concurrence with meat-packing plants.

4.13.3.7.2 Inedible Rendering Process

It involves formation of tallow and grease in concurrence with inedible plants. This is further classified into dry rendering and wet rendering. In dry rendering material is dehydrated to remove water which in turn releases the fat while as wet rendering the fresh material is boiled in water to obtain the fat.

4.13.3.8 BIOREMEDIATION

It is an offshoot of biotechnology that involves the aid of living organisms, like microbes and bacteria withdrawal of pollutants, toxins, and contaminants from water environment and soil. It is mostly used for cleaning of groundwater. This technique is increasing day by day and has got more area of interest (Amin et al., 2012). It is a naturally occurring process where microbes transform or immobilize the contaminants present in an environment into safe end products (Thassitou and Arvanitoyannis, 2001).

4.13.3.9 CONDITIONING

This improves the properties of solids for further processing. The various novel technologies are applied for reason of conditioning involve:

4.13.3.9.1 Ultra Sound Degradation

Before the onset of the digestion process, the material which is in the form of solid waste is put through sound waves so that high pressure and high temperature is achieved with solid matter. As a result of these microbial cells gets broken down because of induced shear stress which in turn results in discharge of gas bubble. In this technique, the rate of disruption of cell increases with fall in AD time. The quantity of sludge is reduced with increase in biogas formation.

4.13.3.9.2 Enzyme Conditioning

Nutrient blends (humic acids, amino acids), enzymes, aerobic, and anaerobic bacterial cultures are applied to digestion systems in this technique. Enzymes are formed by microbes which result in the breakdown of organic matter and converting them into CO_2 and water. This approach is considered to increase the dewaterability of the solids thus by reducing the odors throughout t the digestion process.

4.13.3.9.3 Biological Cell Destruction

In this process, the chemical agent is added to a vessel to activate the sludge so as to divert it from settling tank. The cell wall of bacteria gets weakened by the chemical agent due to the strong oxidation characteristics.

When they decompose into CO_2 and water, the bacteria are sent back to the activated sludge.

4.13.3.9.4 Chemical Cell Destruction

The primary scope is to improve the AD and amount of biogas by disrupting the cell membrane of microorganisms in waste activated sludge process. In the process, caustic soda is added to the waste material and treatment is carried for about one hour. The cell membrane of microorganism becomes weak, and hence the viscosity of solution reduces.

4.13.3.10 ELECTROCOAGULATION

The word electrocoagulation is derived from two words electro means electrical change applied to water and coagulation means the change of charge on the particle surface in order to form an agglomeration. In this process the rate of AD of waste material gets increased. The process implies electric current that breakdown anode and chemically reactive aluminum is dispersed into the wastewater stream. The negatively charged ions allure towards the positively charged aluminum charged ions which result in agglomeration. Due to this, the agglomerated particle size gets increased, hence their weight also increases, which results in settling down of particles.

4.14 AGRICULTURAL WASTE AND ITS MANAGEMENT

Agricultural waste is the by-product obtained after processing of different agricultural products. It consists of numerous wastes which may include animal waste, food processing waste, and crop waste. In other words, it can be defined as the undesirable waste formed as a result of agricultural venture. Agricultural WM is a technique or practice of properly handling the by-products/unwanted materials obtained after processing different agricultural products. AWM provides a healthy environment and can minimize the dire need for fertilizers. It is a planned system which protects the environment from hazardous water and air pollution.

4.15 NOVEL TECHNIQUES FOR AGRICULTURAL MANAGEMENT

4.15.1 PYROLYSIS

In this system temperature of about 400–600°C is given to the waste to heat it up in the absence of oxygen so that portion of the material is vaporized with the charas residue. This is considered to be the best technique used for utilization of agricultural wastes. This technique is used to produce oil, char, and low heating value gases.

4.15.2 ANAEROBIC DIGESTION

Mostly, it is used for methane gas production. It involves two steps in the methane formation from agricultural waste by microbial fermentation: first, the volatile solids are broken down into organic acids by acid-forming bacteria, further the organic acids are acted upon by methanogenic microbe to give methane gas. The important feature about this technique is that it encapsulates the odor and hence the treated waste becomes odor free. It has high retention fertilizer value.

4.15.3 HYDRO GASIFICATION

The process in which hydrogen reacts with the coal at higher pressure and temperature so that carbon present in the coal reacts to form methane is called hydro gasification. In this process, the waste is subjected to high pressure and temperature in reactor, where the waste is converted into useful methane gas.

4.15.4 HYDROLYSIS

Hydrolysis is derived from the Greek word hydro means water and lysis means unbind. It is the process where water molecule breaks one or more chemical bonds. In this process, the agricultural waste is subjected to enzymatic hydrolysis where enzymes come from fungi, these enzymes act on this waste and convert it into useful products.

4.15.5 VERMICULTURE

It is a technique of preparing earthworms feeding on waste material and soil. This technique has replaced conventional composting process (Roupas et al., 2007). The nutrients are increased in soil by releasing the digested material back to the soil (Singh et al., 2008).

4.16 CONCLUSION

Agricultural waste is a residue from the cultivation and processing of raw agricultural materials, which are non-product production and processing outputs and can contain material that is beneficial to humans. These residues are derived from a variety of agricultural activities, including planting, production of livestock and aquaculture. Such waste can be converted into useful materials for human and agricultural use when properly handled through the implementation of innovative methods and AWMSs such as the 3Rs. It is essential that proper waste collection, storage, care, transition, and use are panacea for a healthy world, not from the results so far. Proper waste use will help to grow our agricultural sector and will provide many with viable biofuel options.

KEYWORDS

- agricultural waste
- biofuel
- bioremediation
- pyrolysis
- vermiculture

REFERENCES

Anupam, P., Ankita, H. T., Priyanka, H., Satish, C. P., Ashutosh, P., & Tushar, J., (2019). Various approaches for food waste processing and its management. In: *Global Initiatives for Waste Reduction and Cutting Food Loss*, 191.

Buzby, J. C., & Hyman, J., (2012). "Total and per capita value of food loss in the United States." *Food Policy, 37*(5), 561–570.

Chowdhury, A. H., Mohammad, N., Haque, M. R. U., & Hossain, T., (2014). Developing 3Rs (reduce, reuse, and recycle) strategy for waste management in the urban areas of Bangladesh: Socioeconomic and climate adoption mitigation option. *IOSR Journal of Environmental Science, Toxicology, and Food Technology, 8*(5), 9–18.

Conly, L. H., & Dae, Y. C., (2019). Agricultural waste management in: Food processing. *Handbook of Farm, Dairy, and Food Machinery Engineering.* Myer Kutz Associates, Inc., Delmar, New York.

Esparza, I., Jiménez-Moreno, N., Bimbela, F., Ancín-Azpilicueta, C., & Gandía, L. M., (2020). Fruit and vegetable waste management: Conventional and emerging approaches. *Journal of Environmental Management, 265.*

FAO, (2011). *Global Food Losses and Food Waste – Extent, Causes, and Prevention.* Rome.

Filimonau, V., & De Coteau, D. A., (2019). Food waste management in hospitality operations: A critical review. *Tourism Management, 71*, 234–245.

Filimonau, V., Zhang, H., & Wang, L., (2020c). Food waste management in Shanghai full-service restaurants: A senior managers' perspective. *J. Clean. Prod., 258.*

Jaspal, S., Richa, S., Vandana, B., & Anita, S., (2018). The importance of waste management to environmental sanitation: A review. *Advances in Bioresearch, 2*, 202–207.

Kasavan, S., Mohamed, A. F., & Halim, S. A., (2019). Drivers of food waste generation: Case study of island-based hotels in Langkawi, Malaysia. *Waste Management, 91*, 72–79.

Kunwar, P., Sandeep, K. K., Monika, Y., Nidhi, P., Aakash, C., & Vivekanand, V., (2017). Food waste to energy: An overview of sustainable approaches for food waste management and nutrient recycling. *Biomed Res. Int.,* 1–19.

Li, Y., Park, S. Y., & Zhu, J., (2011). Solid-state anaerobic digestion for methane production from organic waste. *Renew. Sustain. Energy Rev., 15*, 821–826.

Morita, M., & Sasaki, K., (2012). Factors influencing the degradation of garbage in methanogenic bioreactors and impacts on biogas formation. *Applied Microbiology and Biotechnology, 94*(3), 575–582.

Nasir, I. M., Ghazi, T. I. M., & Omar, R., (2012). Production of biogas from solid organic wastes through anaerobic digestion: A review. *Applied Microbiology and Biotechnology, 95*(2), 321–329.

Ostergren, K., Gustavsson, J., Bos-Brouwers, H., Timmermans, T., Hansen, O. J., Møller, H., & Redlingshöfer, B., (2014). *FUSIONS Definitional Framework for Food Waste.* Full Report.

Papargyropoulou, E., Steinberger, J., Wright, N., Lozano, R., Padfield, R., & Ujang, Z., (2019). Patterns and causes of food waste in the hospitality and foodservice sector: Food waste prevention insights from Malaysia. *Sustainability, 11*, 6016.

Parfitt, J., Barthel, M., & Macnaughton, S., (2010). Food waste within food supply chains: Quantification and potential for change to 2050. *Phil. Trans. R. Soc., 365*(1554), 3065–3081.

Plazzotta, S., Manzocco, L., & Nicoli, M. C., (2017). Fruit and vegetable waste management and the challenge of fresh-cut salad. *Trends in Food Science & Technology, 63*, 51–59.

Prabhash, K. D., (2017). *World Food Day: Why 19 Crore Indians go to Bed Hungry Every Night.* India today magazine. URL: https://www.indiatoday.in/amp/india/story (accessed on 3 September 2022).

Pran, K. B., & Singh, R. P., (2018). The effects of toxic agricultural wastes on the environment and their management. *Cosmos, 7*(1), 2319–8966.

Purabi, R. G., Derek, F., Shashi, B. S., & Gerrard, E. J. P., (2016). Progress towards sustainable utilization and management of food wastes in the global economy. *International Journal of Food Science.*

Redlingshofer, B., & Soyeux, A., (2012). "Food losses and wastage as a sustainability indicator of food and farming systems." In: *Proceedings of the Producing and Reproducing Farming Systems: New Modes of Organization for Sustainable Food Systems of Tomorrow, 10th European IFSA Symposium.* Aarhus, Denmark.

Roupas, P., Silva, K. D., Smithers, G., Ferguson, A., & Alamanda, (2007). Waste management co-product recovery in red and white meat processing. In: *Handbook of Waste Management and Co-Product Recovery in Food Processing* (pp. 305–331). New York, CRC Press.

Salihoglu, G., Salihoglu, N. K., Ucaroglu, S., & Banar, M., (2018). Food loss and waste management in Turkey. *Bioresource Technology, 248*, 88–99.

Segre, A., & Gaiani, S., (2011). *Transforming Food Waste into Resource.* Cambridge, UK: Royal Society of Chemistry.

Sindhu, R., Gnansounou, E., Rebello, S., Binod, P., Varjani, S., Thakur, I. S., & Pandey, A., (2019). Conversion of food and kitchen waste to value-added products. *Journal of Environmental Management, 241*, 619–630.

Singh, J., Saxena, R., Bharti, V., & Singh, A., (2018). The importance of waste management to environmental sanitation: A review. Adv. *Biores., 9*(2), 202–207.

Singh, K., Bhimawat, B., & Punjabi, N., (2008). Adoption of vermiculture technology by tribal farmers in Udaipur district of Rajasthan. *International Journal of Rural Studies*, 1–3.

Tawheed, A., Poonam, C., & Suman, V. B., (2012). Recent advancements in food waste management *International Journal of Advancements in Research & Technology, 1*(4), 1–7.

Thassitou, P., & Arvanitoyannis, I., (2001). "Bioremediation: A novel approach to food waste management." *Trends in Food Science & Technology*, 185–196.

UgoDe, C., (2020). Agricultural waste recycling in horticultural intensive farming systems by on-farm composting and compost-based tea application improves soil quality and plant health: A review under the perspective of a circular economy *Science of The Total Environment, 738.*

USDA, (2013). Agricultural waste concept, generation, utilization, and management. In: *Agricultural Waste Management Field Handbook.* United States Department of Agriculture, Soil conservation Service.

Uzmi, A., (2018). *As Millions Go Hungry, India Eyes Ways to Stop Wasting $15 Billion of Food a Year.* Reuters. URL: https://www.reuters.com/article/india-food-hunger-idINKBN1EU0UM (accessed on 30 December 2021).

Wang, L., Liu, G., Liu, X., Liu, Y., Gao, J., Zhou, B., Gao, S., & Cheng, S., (2017). The weigh of unfinished plate: A survey-based characterization of restaurant food waste in Chinese cities. *Waste Manag., 66,* 3–12.

Zhongzhong, W., Yan, J., Shun, W., Yizhen, Z., Yuasheng, H., Zhen-Hu-Hu, Guangxue, W., & Xinmin, Z., (2020). Impact of total solids content on anaerobic co-digestion of pig manure and food waste: Insights into shifting of the methanogenic pathway. *Waste Management, 114,* 96–106.

CHAPTER 5

Food Waste Management: Approaches to Achieve Food Security and Sustainability

AFNAN ASHRAF,[1] SYED ANAM UL HAQ,[2] SHABIR HASSAN,[3] SHAKEEL AHMAD BHAT,[4] SHAHID QAYOOM,[5] ABBAS AHMAD MIR,[6] and MAHSA MIRZAKHANI[7]

[1] *Department of Soil and Water Engineering, Punjab Agricultural University, Ludhiana, Punjab, India*

[2] *Department of Plant Biotechnology, Sher-e-Kashmir University of Agricultural Sciences and Technology, Shalimar, Srinagar, Jammu and Kashmir, India*

[3] *Department of Medicine, Harvard Medical School, Boston, United States*

[4] *College of Agricultural Engineering, Sher-e-Kashmir University of Agricultural Sciences and Technology, Shalimar, Srinagar, Jammu and Kashmir, India, E-mail: wakeelbhat@gmaill.com*

[5] *Department of Fruit Science, Sheri-Kashmir University of Agricultural Sciences and Technology, Shalimar, Srinagar, Jammu and Kashmir, India*

[6] *Department of Geography and Disaster Management, University of Kashmir, Jammu and Kashmir, India*

[7] *Department of Natural Resources, Isfahan, University of Technology, Iran*

Integrated Waste Management Approaches for Food and Agricultural Byproducts. Tawheed Amin, PhD, Omar Bashir, Shakeel Ahmad Bhat & Muneeb Ahmad Malik (Eds.)
© 2023 Apple Academic Press, Inc. Co-published with CRC Press (Taylor & Francis)

ABSTRACT

Food waste management (WM) is one of the prime concerns related to food security and sustainability. Eradication of hunger is one of the important goals of the United Nations sustainable development program. However, the ever-increasing population across the world is hindering the achievement of this goal since the population is expected to be 10 billion by 2050. Not only this, the worldwide problem of food wastage is taking its toll on sustainability, food security, and food safety. Food losses and waste indirectly cost the environment since food production is resource-intensive, which consequently leads to environmental stress, causing water and air pollution, degradation, depletion of soil and greenhouse gases (GHGs) produced through food production, transportation, conveyance, and WM processes. As a result of these growing pressures, social, and economic issues regarding food waste are being recognized as a core problem for states, scholars, non-governmental organizations, companies, and the general public. Therefore, this chapter deals with the different approaches of sustainable and secure management of food waste that can definitely help us secure the present as well as the future to tackle the existing crisis of food all across the globe.

5.1 INTRODUCTION

Food waste is generally described as the loss of materials intended for consumption by humans that are either discharged, polluted, degraded or eventually lost. As per United Nations Food and Agriculture Organization (FAO), food loss (FL), or waste is "any change in the accessibility, edibility, wholesomeness or consistency of eatable material that prevents people from consuming it." This description was set out for the time after the end of the food harvest, where the argument is that the final customer is the proprietor (FAO, 1981). Another FL concept put forth by Gustavsson, Cederberg, Sonesson, van Otterdijk, and Meybeck (2011) provided a description of the stage of development of the food supply chain (FSC) along with the post-harvest and processing step. According to Parfitt, Barthel, and Macnaughton (2010), "Food waste is the shortage of food arising either at the retail level or at the final stage of consumption, and is created by negligent actions of both retailers and consumers." According to Gunders (2012), more than 30% of edible food is wasted annually in the United States, mainly in households, restaurants, and foodservice establishments. In Europe, about 89 million tons of food is lost annually, which accounts for 180 kg of food per capita, but this number does not include losses during food processing and harvest processes. The sum wasted per head per year was found to be: 110, 109, 108, 99, 82 and 72 kgs in the United Kingdom, US, Italy, France, Germany, and Sweden respectively, when looking only at household waste (Martínez, Menacho, and Pachón-Ariza, 2014). Private households contribute the greatest fraction of food waste to the food supply system (Parfitt et al., 2010). Major losses are sustained at the initial stage of the FSC in developing nations, primarily due to the limitation of planting, processing, and recycling methods or the lack of sufficient transport and storage facilities. Developed countries contribute highest towards food wastage during the end steps of the operation. However, the losses reported at the farm level are not negligible even in these countries. For example, in Italy, about 17.70 million tons of agricultural food commodities were abandoned in the fields in 2009, which account to 3.25% of total production (Segrè and Gaiani, 2012).

Food scarcity is due to temperature and environmental dynamics, as well as pathogens and parasites. But when a comparison is made between developing and advanced countries having access to technology, agricultural expertise and way-outs used to manage agricultural lands, infrastructure, sowing, planting, harvesting, processing, and storage, there are differences at this point. Regulatory and economic considerations are helpful in developing countries, and often even in emerging countries. However, certainly there is still also much to do to grasp the reasons for the shortages in the FSCs at the early stages. Inefficiencies and technological malfunctions in manufacturing processes are the key reasons for the waste generated during the early preparation phase of the agricultural products and the semi-finished goods – typically referred to as 'production waste.' According to FAO statistics, based on 2011 Food Balance Sheets, the yearly amount of food waste produced globally has a carbon footprint of 3.6 Gigatons of CO_2 equivalents (Scialabba, 2015). The yearly effect of food waste is 186 Metric Tons (Mt) CO_2 Equivalent. This represents about 16% of the total of the European Union (EU)-level carbon footprint of the global food chain (Scherhaufer et al., 2018). According to the FAO's (2009) product prices, the economic cost of the global surplus of farm food products, with the exception of fish and shrimp, is 750 billion USD at the consumer level (Food, 2013). In 2014, FAO revised forecasts of prices for 2012 and substituted producer prices for import/export consumer prices for post-agricultural wastage. This leads to an overall numerical value of 936 billion dollars of global food waste (Footprint, 2014). The cost of edible waste at the European level for the EU-28 in 2012 is estimated at approximately 143 billion dollars, depending on the amount of food intended for human consumption at each point across the FSC where it is destroyed (Fusions, 2016).

5.2 SUSTAINABILITY OF FOOD SECURITY: A CHALLENGE AND APOLITICAL IMPERATIVE

Sustainability of food security demands: (a) access to or adequate production of food; (b) accessible and purchasable food; (c) adequacy of nutrients, including resources, proteins, and micronutrients, and safety; and (d) the consistency along with predictability of these requirements (Helland and Sørbø, 2014). Risks to food protection comprise of inadequate availability of healthier food or reduced consumer buying power (Bazerghi, McKay, and Dunn, 2016). Food poverty is most impacted by low-income populations with higher hunger and malnutrition threats. A link between food prices and social unrest in low-income nations was found by the International Monetary Fund (Arezki and Bruckner, 2011). As a consequence, socio-economic stress caused malnutrition and food poverty (Helland and Sørbø, 2014). Food costs are increasing amid the breakdown of political institutions, social security networks, population tensions and the existence of other grievances connected with social instability. An example of increasing food prices as a cause of social instability is the Arab Spring 2011 (Johnstone and Mazo, 2011). Consequently, food poverty represents a big issue in the fields of healthcare, social development and political harmony. The population is estimated to go higher by 2050 to around 10 billion. A diet rich in animal protein will be eaten by those 10 billion people (Sundström et al., 2014). The biggest cause for this is diet rich in meat, fish, poultry, and milk products would be used by 3 billion people. It is also probable that sustainably feeding 10 billion people within the next 20 years will mean disruptive improvements in the FSCs (Gerhardt et al., 2020).

One more concern becomes conversion of agricultural crops such as; corn, sugarcane, etc., into biofuel production. Around 400 million people would have been benefitted with these biofuel-diverted crops (Helland and Sørbø, 2014). Furthermore, these conversions make biofuel crop prices more related to the price of gasoline than to the availability and demand of food. Growing food price speculation is another concern as hedge funds become more important agents in food crop markets (e.g., wheat, oilseeds, and maize). Rising food price risks pose a threat to food security for poor communities. Growing urbanization has forced a large proportion of the world's population to live in villages, one more food security issue. The food security of low-income urban communities experiences higher food prices or sudden increases in food prices, particularly for staple foods, in countries with insufficient socio-economic safety nets, as urban consumers rely on the opportunity to buy food. Moreover, the remedy may not be based on imports, since certain nations, in periods of inadequate food supply, enforce export restrictions on food. Long-term food security solutions are also national goals that will describe the push for the development of innovative food and animal feed sources and resilient grocery chains. Measures to ensure food safety and consistency can, however, often limit the quantity of food available, thus amplifying food shortages. The key argument is that food security and protection, is part of socio-economic sustainability. The attainment of one objective may violate other objectives. Therefore, reconciliation of both priorities and negotiate the trade-offs needs to be done.

5.3 FOOD LOSSES AND WASTE VERSUS FOOD SECURITY

As per Food and Agricultural Organization (FAO), FL includes the supplies that we lose between producer and consumer along the food chain, while food waste discards healthy and nutritive foods (Losses and Waste, 2011). According to a report by Unilever in 2019, there is a loss of about 1/3rd of the total food produced globally. Nearly 28% of the global agriculture area and 8% of global greenhouse emissions are accounted for by lost and unused food. In addition, the elimination of food waste and loss will generate a major market potential valued approximately more than 400 billion US Dollars (Vågsholm, Arzoomand, and Boqvist, 2020). Food wastage often reflect the wasted labor, money, water, electricity, land, and other resources that have been used to produce food and thus endanger sustainability. There are essential ties between sustainability and food wastage. In the FSC, more than a quarter of the total food produced is lost (Lipinski et al., 2013). Such wasted food accounts for nearly 24% of the overall energy content of the food produced, demonstrating tremendous opportunities for improved food security. Minimizing the waste and depletion of global food could feed more than a billion new people. More efficient use of land and water resources with positive impacts on the environment and livelihoods would benefit from less food wastage.

The FL issue can be illustrated by the world's fish farming. Fish farming/processing accounts for about 20% of total human protein intake (Moffitt and Cajas-Cano, 2014). During the farming process, 20 to 30% of catches are lost (Ye et al., 2017), whilst the other 10 to 15% of catches are allocated as feed to fishes, leading to protein loss. Another example is that nearly half (46%) of the overall grain and vegetable harvest is used as animal feed, while only about 1/3rd is used to feed human beings. During this conversion process of plant to animal feed, much of the nutrition is lost. The likelihood of increasing traditional cereal-based meat, milk, and egg production is limited by the lack of adequate cultivable land and water, and also the challenge of increasing agricultural and animal production (Gerhardt et al., 2020). In this context, if the production of animal food were doubled, it would mean an expansion of the production of cereals and vegetables above sustainable levels of production (Gerhardt et al., 2020).

5.4 CAUSES OF FOOD WASTAGE

The causes of food wastage include:

- Faulty agricultural practices, harvesting, packaging, distribution, and marketing and consequent spoilage.
- Untoward reduction in the consistency of packages, fruits or vegetables, such as bruising.
- Quality defects, such as fresh products that deviate from desirable shape and color, such as curvy cucumbers, discarding inappropriately sized apples while grading operations.
- Retailers and customers recycle foods that are similar to, at, or past the "best-before" dates.

- Significant amounts of nutritious edible food are frequently leftover from households, restaurants, and finally discarded.
- Food recall from the manufacturer or store.

For example, these recalled foods (e.g., nuts) can pose a danger to a minority of the population due to the presence of allergens. For most of the population that are not allergic to nuts, the recalled foods would be healthy. Unspecific product recalls raise the waste of food. In the United States, 42% of the 382 recalls were found to be attributable to unreported allergens, 18% to the existence of foodborne pathogens and 7% to the occurrence of foreign materials (Vågsholm et al., 2020).

5.5 ASSESSING THE WAY OUTS FOR REDUCING FOOD WASTAGE

A hierarchy of food wastage reducing options has been identified by the U.S. Environmental Protection Agency (Ceryes et al., 2021) (Figure 5.1). Mourad (2016), performed comparison of policies of France and the United States towards food waste reduction. The results showed the three contradictory orders of alternatives to leftover food supply, based on natural, monetary, and societal justifications. Implementation of efficient food waste reduction techniques, for example, altering the approval standards for fresh yield rather than feeble prevention approaches (creating novel practices for food corporate workers), possibly will do more in terms of lasting sustainability. Although the goals of curtailing FL and waste are exceedingly advantageous thus, this pursuit should involve food safety. Tradeoff's, judgmental actions made, should be based on facts and justified in a straightforward manner.

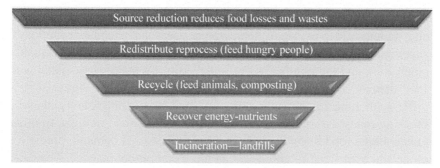

FIGURE 5.1 The food recovery hierarchy (Ceryes et al., 2021).

5.5.1 SOURCE REDUCTION

5.5.1.1 SOURCE REDUCTION – REDUCE WASTE AND LOSSES

We need new approaches to look at the safety and preservation of food. The result of this change in viewpoint is that food policy should lay emphasis not on food production but on reaching zero hunger and healthy nutrition. Novel ways of thinking about sustainability and food could be opened up by a shift of perspective. The premium should be charged to make sure that one does not suffer from malnutrition, get adequate nutrition and avoid foodborne diseases in future, if food security is seen as an insurance concern (Lovins, 1990). Investment in reducing the quantity of food lost or discarded (reducing the source) would be as important as investment in increased capacity to manufacture food. A main component of the future strategy to sustainably feed 10 billion people would also be to reduce the cause of food depletion and pollution from farm to fork.

5.5.1.2 SOURCE REDUCTION—USING NOVEL INFORMATION TECHNOLOGY AND ARTIFICIAL INTELLIGENCE SOLUTIONS

We have to eliminate FL and waste in order to achieve sustainability. This plan goes well with the resource and environmental footprint reduction objectives. Conversely, consumers want to buy their favorite food when it's convenient. Food producers face trouble in getting the inventory practices correct when reordering food and maintaining the required stock levels. In other words, the simultaneous problems of reducing food waste and stock-outs are a complex resource allocation problem for food businesses and retailers. Food waste and stock-outs are primarily attributed to poor market forecasting and incorrect ordering of goods as a result (Arunraj and Ahrens, 2015). Prices, temperature, seasons, events or holidays, supermarket or competing industry sales or discounts, product quality (shelf life), and the number of customer visits are factors associated with food demand in retail stores. In addition, high volatility and skewness differ with time in the time series of food retail transactions, thereby breaking many assumptions in the traditional statistical models. In order to encourage food business operators to forecast food demand, predictive models like artificial intelligence can more commonly be used. The daily sales of perishable commodities could be reliably measured with the use of seasonal autoregressive models of synchronized moving averages with external variables (Arunraj and Ahrens, 2015). If two or more models are used with separate analytical techniques, the expected accuracies will improve. For example, the market for fresh food in grocery stores could be best modeled by integrating moving averages with neural backpropagation networks (Chen, Lee, Kuo, Chen, and Chen, 2010). In addition, food sources can be estimated precisely by using big data tools (Figure 5.2), for example, remote sensing data, metrology, photos, genetic records, and historical output data (Gounden, Irvine, and Wood, 2015). In this respect, compared to machine learning approaches, Bayesian networks would be a more straightforward technique that would allow deeper understanding into the FSC. One major concern is that, in order to achieve maximum benefits, food safety and quality aspects need inclusion in these big data models. In the big data era, complementary, and conflicting visions for healthy food exist (Nychas, Panagou, and Mohareb, 2016; Ropodi, Panagou, and Nychas, 2016).

Multidisciplinary methods must include better resources for the food industry to ensure food quality and safety (Ropodi et al., 2016). Complex and not so simple algorithms like deep learning and artificial intelligence have feasible approaches for processing these massive data sets in real-time (Ropodi et al., 2016). The absence of clarity makes it difficult to interpret the findings and give evidence-based advice. Furthermore, a crisis formula may be the novel implementation of food supply and demand forecasting, as well as tracking the food chain from farm to fork, while avoiding food safety concerns. Big data techniques should also include aspects of food safety. The benefits may include reducing food waste, certifying food security through more effective process management, and reassuring consumer's faith in the food industry. In addition, a shift from an invasive or disruptive testing to non-invasive sensor-based automated checking would be helpful. These sensors are mounted easily onsite and can track the performance in real-time. Through these instruments, the immense quantity of peak output, analytical, and imaging metadata assembled will offer a general view of the processes of spoilage and degradation of different food items under various storage conditions (temperature and packaging). The online availability of this growing expertise will give the food industry continuous benefits. By lowering food and waste losses while maintaining food safety and consistency, big data analytics promise a way forward to achieve greater productivity.

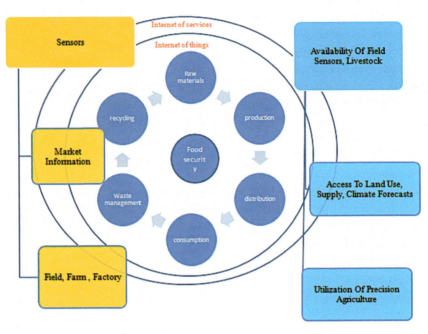

FIGURE 5.2 Big data framework for food security (Irani et al., 2018).

5.5.1.3 SOURCE REDUCTION BY INTELLIGENT LABELING (SHELF LIVES "BEST BEFORE DATE" AND "USE BY DATE")

Food shelf life is either known as "best before dates" or "use-by dates." The "best before date" is a method for the management of food quality. The food business company guarantees that foods eaten before the best day, e.g., cold, dim storage, are of good quality if the storage orders are followed. The food is healthy after the best before the date but the standard can decrease. The customer may consider the food to be unsafe and unhealthy, and is thus discarded. This is because consumers confuse the "best before dates" with the "use-by dates," interpreting them as "poisonous dates thereafter." The "use-by date" is a tool for food safety management, meaning that the food is safe to eat before the "use by date," provided that the food is processed in accordance with the instructions of the food company operators. After the "use by date," the food business operator guarantees food protection. Customers can then discard the food. On the last "use by date," the foodstuff must comply with the microbiological protection end-product requirements. Therefore, the shelf life of a foodstuff would be determined by these criteria. There are major differences in vulnerability between groups that are healthy and immuno-compromised, but the debate is complicated (Rocourt, BenEmbarek, Toyofuku, and Schlundt, 2003), because listeriosis is mostly infectious in people with reduced or altered immune function. Organ transplant individuals, pregnant women and people doing dialysis are nearly 3,000, 1,000 and 500 times more likely to catch listeriosis than normal individuals (Rocourt et al., 2003). The fact that one in seven cases of listeriosis in pregnant women is further raising the burden of the disease (Wadhwa and Smith, 2017).

5.5.1.4   INTELLIGENT LABELING AND SENSORS

Smart packaging and sensor systems possess tremendous potential to minimize food waste and hence enhance food safety (Poyatos-Racionero, Ros-Lis, Vivancos, and Martínez-Máñez, 2018). There are four sensor families, typically gas sensors, time-temperature indicators (TTI), and identification tags for freshness, food package integrity (Poyatos-Racionero et al., 2018). The TTIs are useful to detect temperature abuse along the distribution chain and could allow for complex dating of the shelf life of a foodstuff. The "use-by date" may be expected to be replaced by, for example, the TTI indicator changing red if the food is not appropriate for human intake or yellow if the quality of the food declines. The indicators of freshness track freshness by responding to food-generated metabolites. Organic acids, pH, biogenic amines and CO_2 are such markers. For each foodstuff, the freshness indicators require calibration. For fish products, certain metrics that are acceptable for the freshness of leafy greens may be unsuitable. Gas sensors can test the consistency of food packaging, the leakage of the protective atmosphere, or shifts in the metabolites of gas (Ghaani, Cozzolino, Castelli, and Farris, 2016). To make sensors a commercially feasible, they should be of low cost, reusable, and long-lasting. Educating customers about the value of sensor outputs will possibly be more complicated. Radiofrequency identification detectors (RFID) tags help to track and track a food product as they provide real-time information on the identity of the food product. In addition, RFID tags could provide data to analyze the causes of waste and loss of food. Controlling food theft may be an added advantage. In recent years, food fraud and fraudulent contributions to food production have emerged as a concern (Manning and Soon, 2016). The sources and destiny of food in foods that are vulnerable to theft could be confirmed by RFID tags combined with blockchains. Blockchains are resistant to tampering and fraud and can promote secure exchange and minimize paperwork (Ahmed and Ten Broek, 2017). While using blockchains, the traceback can be easier and more accurate when investigating outbreaks. Thus, one may specifically target food recalls. The capacity of governments to alleviate drought by imports of food from other areas is an essential aspect of food security. Faith in food safety is crucial, and the application of blockchains may promote confidence and enhance procedures.

5.5.1.5 SOURCE REDUCTION – REDUCE FOOD LOSSES BY INTENSIFICATION OF FOOD PRODUCTION

Food production intensification needs to be compatible with long term sustainability criteria (Rockström et al., 2017). For long-term sustainable agriculture, it must function within its environmental limits. The main criteria include ecological aspects, resource footprints and resilience, social dimensions and resilience.

The production of food for agriculture and aquaculture should transform from a catalyst for global environmental change to a basis for global biodiversity. In addition, the resilience of food security or the capacity to cope with food production and distribution stresses without raising the risk of foodborne diseases, malnutrition or hunger, is essential. Smart intensification of food production would need the double pressure of reducing arable land and growing global population (Gerhardt et al., 2020). More than half of the production of edible vegetables is wasted in the fields (Johnson et al., 2018). The losses are due to the failure in the harvesting process, the failure to comply with the requirements of the supermarket. This demonstrates an easy way to increase the quantity of food supplied and to boost food security with minimal downside risks. In addition, sustainability would increase because doubling the quantities of plant foods available to customers' needs the same resource footprints, i.e., fertilizer, labor, irrigation water, and chemical inputs. Another solution may be to turn to novel animal proteins extracted from food waste and by-products fed by insects replacing conventional meat products.

By consuming cereals intended for animal feedstuffs, a third path for enhancing resilience is moving down the food chain. If animal production is focused on feed that is unavailable for human consumption, i.e., pastures, and grasslands, sustainability, and food security will improve. This will suggest that the pastoral production of mutton and milk will make a significant offering to the supply of proteins of good quality in order to ensure potential food welfare. Increased livestock production to boost animal welfare, health, and disease prevention would be a very significant contributor to sustainable development in countries with low and medium incomes (McDermott, Staal, Freeman, Herrero, and Van de Steeg, 2010).

5.5.2 REDISTRIBUTE OR REPROCESS FOOD

By way of food banks and food donation services, one can redistribute food (Schneider, 2013). For making food accessible for deprived socio-economic groups, such steps may be necessary. The donated foods are usually near the end of their shelf life, which may suggest an elevated risk of foodborne illness or "best before dates" of potential reduced quality in "use-by dates."

5.5.2.1 FOOD DONATIONS—LIABILITY CONCERNS

Austrian supermarkets, for instance, waste about 10% of bread (Lebersorger and Schneider, 2014). Just 7% of the surplus bread is contributed to food banks, i.e., less than 1% of overall bread supply. The concerns of food business owners over their duty to donate perishable goods that could be related to foodborne illness are one cause for this. The USA and Italy have introduced Good Samaritan Laws to ease certain liability concerns. These statutes shield beneficiaries from responsibility when giving to a non-profit group and from civil and criminal liability when good contributions in good faith unintentionally harm one of the recipients in need (von Braun, 2010). Legislation about how to donate food without incurring liabilities would be needed to support food donations and food banks.

5.5.2.2 FOOD DONATIONS—RESILIENCE AND NOVEL INFORMATION TECHNOLOGY SOLUTIONS

In order to minimize food insecurity, food donation programs and food banks may help (Authority, 2006). They will prevent malnutrition and thereby, increase healthcare for disadvantaged socio-economic groups if these programs are well planned and controlled. This would enhance the resiliency and stability of a society. Non-perishable goods, including canned, frozen, and dry foods, are common gifts to food banks. Food banks can also be found in the form of ambulance services, such as ambulance centers, soup kitchens, and food pantries. To alleviate severe food deprivation and the risk of starvation, food banks are critical (Bazerghi et al., 2016). Apart from this, because of the restricted availability of nutrient-dense and perishable foods may possess lesser options to enhance the nutritional status of the recipient. However, if food banks fulfill the nutritional needs of their customers and the supply of perishable food is sufficient, food security contributions will be important. Moreover, for transportation and food storage, food banks typically have limited economic resources. The biggest obstacle for food banks is predicting their stock of food, which impairs their capacity to transport, preserve, and allocate donated food cost-effectively (Brock III and Davis, 2015). In addition, to provide their members with a healthy diet, food banks also need to buy alternative ingredients. For predicting the dynamics of food sources, novel IT and artificial intelligence solutions appear ideally suited.

5.5.2.3 FOOD DONATIONS AND FOOD SAFETY

Food donation chains are less structured than conventional food chains and have imperfect cold chains (De Boeck, Jacxsens, Goubert, and Uyttendaele, 2017). A further situation occurs with the occasional disappearance of food safety instruction by those working on food donations. Perishable foods donated are often available to eat or prepare for meals to be reheated. When 72 samples of perishable foods from Belgian social foods were tested, growing total of listeria monocytogenes (log 3.5 CFU/g) and enterobacteria (6.7 CFU/g) is ready to eat cooked meat were found (Wooldridge, Hartnett, Cox, and Seaman, 2006). It illustrates the need for adequate cooling power if perishable food is to be removed from food banks. Another issue is that, irrespective of other co-morbidities or health conditions, persons consuming donated foods are far more vulnerable to developing foodborne diseases, suggesting the need for cautious trade-offs between food safety and the reduction of food insecurity. However, such programs are a must for the preservation of food at work.

5.5.2.4 REPROCESSING FOODSTUFFS FOR HUMAN CONSUMPTION

Reprocessing also means reheating or frozen foods that go past their "best before dates" or "use-by dates." Salmon filets and beef pieces, for example, are ground into salmon or beef patties and after that cooked for a different shelf life. One more option is freezing foods only before or using them for dates. Using food leftovers as raw materials or reheating the leftovers is a third choice for future meals. An analysis of Swedish supermarkets (Lagerberg Fogelberg, Vågsholm, and Birgersson, 2011) showed a payment proposal for the supermarket was to have its own chef make warm lunches or dinner portions. The amounts of food waste have been decreased, refined into hot lunches, pates, or desserts, and ready-to-eat sandwiches as foods that reach the 'best before date' or with certain noticeable signs associated with less approval by the customer. Moreover, customers offered decent treatment, and there was another revenue opportunity for the supermarkets. Foods meant for the bin tended to be transformed into meals to be sold to boost profits. In fact, the supermarkets found that among the other workers, the recruiting of a chef improved the hygiene standards. Reheating dishes that have already been cooked improves the availability of food, but involves a food safety threat. The development of toxin-containing bacteria associated with slow cooling and subsequent reheating of pea soup is a common example (Rosengren, 2017). In Sweden, almost 20% of confirmed cases of foodborne disease were infected with toxin-producing bacteria-*Staphylococcus aureus*, *Bacillus cereus*, etc. (Vågsholm et al., 2020), while 18.5% of bacterial toxins were associated with a comparable proportion of foodborne outbreaks in the EU in 2018. European Food Safety Authority (EFSA) reported *Staphylococcus aureus* food poisoning is synonymous with cross-contamination and resulting abuse of foodstuff stored at temperatures ranging between 5–60°C (Hennekinne, De Buyser, and Dragacci, 2012). The latter two species are spore formation and the spores survive cooking or associated heat treatments. While food reprocessing is advantageous from the perspective of food security and sustainability, negative trade-offs in food safety must be avoided. It is also a clear recommendation to train food business operators involved in food reprocessing.

5.5.3 RECYCLE (REPROCESS WASTED AND LOST FOOD TO ANIMAL FEED)

As fewer issues about the propagation of animal diseases or zoonotic infections apply, for missing or discarded plant material, recycling, and reprocessing of food as animal feed may be simpler. The bovine spongiform encephalopathy (BSE) epidemic has led to tighter restrictions in the European Union (EU) by feed producers reprocessing plant food and the prohibition of feeding scraps to feed animals. All feed manufacturers processing plant-based food waste for animal feed must be officially monitored. The negative effect of these bans came out to be as restrictions on circular and thus environmentally humane animal production practices. Consequently, only 3 million tons are recycled as animal feed from almost 100 million tons of food waste (Salemdeeb et al., 2017). There are many advantages from replacing feedstuffs relying on cereals or soybeans for plant-based food scraps regarding resource footprints, production, and concentrations of heavy metals penetrating the food chains. In contrast, farmer's prices of feeding recycled food to their livestock are more stable than those of feedstuffs, where the dynamic market dictates the rates. In conclusion, having this trade-off right could unlock major advantages and profits, respectively, for customers and suppliers.

5.5.3.1 CIRCULAR FOOD PRODUCTION—AQUACULTURE AND AQUAPONICS

Circular food systems are very good when it comes to environmental footprints. Circular systems mean that most nutrients are recycled, but ample barriers, like species barriers in this nutrient cycle can lead to a variety of diseases. This means that careful attention should be paid to the biosecurity and potential hurdles of circular food production systems. Combining aquaculture and aquaponics is an example of a circular food production system (Monsees, Kloas, and Wuertz, 2017). Compared to animal agriculture on land, aquaculture has environmental advantages, such as lesser water footprints and greater feed conversion of about 1 kg of feed to 1 kg of fish meat. Since fertilizing plants are the waste produced by fish, a food production system focused on hybrid aquaponics and aquaculture will minimize environmental footprints. Circular food processing systems, however, have problems of their own. The metabolism of the fish produces ammonia that is converted to nitrates in the bio-filters. This change of ammonia to nitrates demands that the pH above 7, subjecting the fish to suboptimal conditions of rearing, resulting health and welfare risks. After that, by adding acids to the water after the bio-filter, the pH is reduced to 5–6 for maximizing plant's nutrient absorption when used for irrigation. The decoupling of aquaponics and aquaculture systems may also be advantageous for fish welfare and health reasons. In addition, yields from independently optimized production systems of aquaponics and aquaculture tend to be greater than for coupled systems.

5.5.3.2 CIRCULAR FOOD SYSTEMS—ADVANTAGES AND RISKS

Circular food systems have been able to lower the waste and depletion of food, increasing the resilience of food systems and food safety systems (Jurgilevich et al., 2016). The quality of our food welfare is questioned today because 30–50% of food is discarded or destroyed at various levels of the food chain. This has resulted in higher animal food intake and higher environmental footprints. Smaller environmental and capital footprints and the processing of nutrients, by-products, and food waste would be required for circular food production processes, resulting in fewer food waste and losses. The improved accountability available from smaller and local grocery outlets is an added advantage. The integration of local and seasonal components in supply chains will increase food supply and demand balance and decrease storage and transportation needs. However, there are hazards when transitioning from a linear production method for extractive foods to a circular and recycled one. Slaughterhouses create a great deal of food waste, such as offal, designated hazard material, condemned carcass fragments or condemned whole carcasses. Food protection laws are essential for the approval or condemnation of carcasses, but too persistent condemnation leads to additional food waste

(Arzoomand et al., 2019). The lack of preparation for meat inspectors and slaughterhouse workers is one of the causes of this undue inspection.

5.5.4 RECOVER ENERGY NUTRIENTS

5.5.4.1 RECOVER AS BIOFUEL AND NUTRIENTS

It focuses on recovering energy and use the nutrients from FL and waste as fertilizers, as well as by-products from animal and vegetable food processing. In terms of food security and sustainability, these recovery strategies are less favorable related to strategies discussed previously. But they may be worthwhile as supplements. There are trade-offs between food security, sustainability, and food protection, even here.

5.5.4.2 RECOVER AS BIOGAS AND NUTRIENTS FROM MANURE

Recycling manure from animal agriculture (Leibler, Dalton, Pekosz, Gray, and Silbergeld, 2017) for industrial uses to recycle energy and recover nutrients as fertilizers is one complementary method. Sustainability and food security are a potential win-win scenario, but with risks to food safety. The quantities of animal manure are mostly concentrated in small areas. Therefore, threats of pathogens being released could occur. If the manure is abandoned unprocessed, the avian flu virus could live for up to 600 days in manure (Graiver, Topliff, Kelling, and Bartelt-Hunt, 2009), providing a possible hot spot for transmission or later re-emergence of the disease. After biogas processing, food waste residues may be useful as fertilizers and reduce the resource footprint of food production. Three strategies for extracting nutrients, energy from biowastes and manure (Albihn and Vinnerås, 2007) are composting, anaerobic digestion (AD) and NH_3 treatment, with various advantages and disadvantages. AD, for example, may generate biogas and degrade organic contaminants, whereas costly high-tech equipment is needed on the downside. The choice of treatment methods should be taken on a case-by-case basis, but the key requirements for the management of pathogens in fertilizer processing were the time-temperature profiles and the content of ammonia.

5.5.4.3 RECOVER ENERGY THROUGH BIOFUEL PRODUCTION

Biogas, biodiesel, bioethanol, and biobutanol are biofuels (Tabatabaei, Karimi, Kumar, and Horváth, 2015). The preference of biofuel manufacturing substrates, for example, in agriculture, should be as varied as possible. Losses and waste in the food industry, household waste, and solids from urban wastewater. EFSA, for instance, has reviewed the biosecurity of the production of biodiesel from animal by-products (Hazards, 2015). There were also potential substrates for fats from sewage cleaning plants, cooking, and frying oils. For biodiesel method, the current biosecurity criterion is to reduce the BSE (Bovine Spongiform Encephalopathy) agent infectivity not less than 1 million cycles (log 6) in order to use offal as a substrate for all risk groups. As biodiesel may substitute kerosene for aircraft and diesel for trucks and agricultural machinery, the sustainability gains stem from smaller climate footprints.

Another sustainability caveat is that it should not be possible to use substrates for the processing of biofuels as human food or as animal feed for which the links between oil and food prices (Al-Maadid, Caporale, Spagnolo, and Spagnolo, 2017) are worrisome. Therefore, if oil prices were to escalate, the prices of global staple commodities like cereals and sugar would increase. This may be a very big and crucial problem for social sustainability and food stability as food markets grow more unpredictable as edible crops are diverted to the production of biofuel to alternative for oil.

5.5.5 INCINERATION AND LANDFILLS

The least suitable solutions for food waste and damage control include incineration and landfills. But, from a sustainability perspective, incineration has many advantages. Power and heat incineration is an alternative that could enhance resilience in developed countries with energy shortages and where food waste is deposited in unsanitary landfills (Unaegbu and Baker, 2019). The food that is lacking or wasted is also usable as fuel which will balance the energy needed to produce the electricity necessary locally. The gains are from the incorporation of food waste into a resource of energy and power that is locally available. This would increase the reliability of building conditions and allow a viable plan for cold chains.

160 Integrated Waste Management Approaches for Food and Agricultural Byproducts

Another example from Sweden is that, if cereals were spoiled by farmers or feed mills because of mold, the grains could be incinerated to recover energy (Vågsholm et al., 2020). Benefits include stopping certain moldy grains from entering the food or feed chains and preventing the foods already lost, owing to molds from utilizing renewable energy. Incineration could, therefore, slash oil consumption and carbon footprints. Therefore, incineration may be regarded as the initiation of the advancement in sustainability and security in food systems.

Landfills are not much popular since they are not desirable by most of the dwellers near them. Landfills tend to be the least sustainable food WM system with the highest environmental and WM system. In addition, the threats to biosecurity and food safety are major because landfills where food waste is discarded are usually a dwelling place of vermin (Oivanen, Mikkonen, and Sukura, 2000).

5.6 CONCLUSIONS

In today's world, food waste has become a big concern. Food wastage has become a major crisis in recent years impacting "developed and developing countries" equally. Approximately 1/3 or 1/5 of all the food produced remains unconsumed. In addition, food waste has a strong impact on the overall economy and environment. In the proper organization of waste food, a variety of resolutions may be applied and prioritized in a parallel manner for managing food waste. Circumvention and donation of palatable parts of social services characterize the greatest beneficial alternatives. In industrial processes, for the processing of biofuels or biopolymers, food waste is also employed. Further measures predict nutrient repossession and carbon fixation by composting. Therefore, beneficial strategies, policies, campaigns, and legislative changes are required to occur for achieving the aforementioned objectives.

In addition, the task of combating food waste must take into account the entire supply chain, from food production to food manufacturing and retailing. There is also a need for research to be done about customer behavior and preference patterns. The causes of food waste differ between developed and developing countries, with most losses occurring at an early, post-harvest, and processing stage in developing countries, while significant losses occur in developed countries at retail and usage stages.

Therefore, the studies and policies needed to minimize food waste must be unique to the region.

KEYWORDS

- **food donation**
- **food scarcity**
- **food security**
- **food waste management**
- **food waste recovery**
- **hunger eradication**
- **incineration**
- **sustainability**

REFERENCES

Ahmed, S., & Ten, B. N., (2017). Blockchain could boost food security. *Nature, 550*(7674), 43–45.

Albihn, A., & Vinnerås, B., (2007). Biosecurity and arable use of manure and biowaste—Treatment alternatives. *Livestock Science, 112*(3), 232–239.

Al-Maadid, A., Caporale, G. M., Spagnolo, F., & Spagnolo, N., (2017). Spillovers between food and energy prices and structural breaks. *International Economics, 150*, 1–18.

Arezki, R., & Bruckner, M., (2011). Food prices and political instability. *IMF Working Papers*, 1–22.

Arunraj, N. S., & Ahrens, D., (2015). A hybrid seasonal autoregressive integrated moving average and quantile regression for daily food sales forecasting. *International Journal of Production Economics, 170*, 321–335.

Arzoomand, N., Vågsholm, I., Niskanen, R., Johansson, A., & Comin, A., (2019). Flexible distribution of tasks in meat inspection: A pilot study. *Food Control, 102*, 166–172.

Authority, E. F. S., (2006). Opinion of the Scientific Panel on biological hazards (BIOHAZ) related to the public health risks of feeding farmed animals with ready-to-use dairy products without further treatment. *EFSA Journal, 4*(4), 340.

Bazerghi, C., McKay, F. H., & Dunn, M., (2016). The role of food banks in addressing food insecurity: A systematic review. *Journal of Community Health, 41*(4), 732–740.

Brock, I. L. G., & Davis, L. B., (2015). Estimating available supermarket commodities for food bank collection in the absence of information. *Expert Systems with Applications, 42*(7), 3450–3461.

Ceryes, C. A., Antonacci, C. C., Harvey, S. A., Spiker, M. L., Bickers, A., & Neff, R. A., (2021). "Maybe it's still good?" A qualitative study of factors influencing food waste and application of the EPA food recovery hierarchy in US supermarkets. *Appetite*, 105111.

Chen, C. Y., Lee, W. I., Kuo, H. M., Chen, C. W., & Chen, K. H., (2010). The study of a forecasting sales model for fresh food. *Expert Systems with Applications, 37*(12), 7696–7702.

De Boeck, E., Jacxsens, L., Goubert, H., & Uyttendaele, M., (2017). Ensuring food safety in food donations: Case study of the Belgian donation/acceptation chain. *Food Research International, 100*, 137–149.

Food and Agriculture Organization (FAO), (1981). *Food Loss Prevention in Perishable Crops*. FAO Agricultural Service Bulletin, No. 43; FAO Statistics Division: Rome, Italy.

Food, F., (2013). Agriculture Organization of the United Nations. *Food Wastage Footprint: Impacts on Natural Resources; Summary Report*. Natural Resources Management and Environment Department: Rome, Italy.

Footprint, F. F. W., (2014). *Full-Cost Accounting*. Final Report.

Fusions, E., (2016). *Estimates of European Food Waste Levels*. https://food.ec.europa. eu/safety/food-waste_en#:~:text=In%20the%20EU%2C%20around%2088%20 million%20tonnes%20of,a%20quality%20meal%20every%20second%20day%20 %28Eurostat%2C%202020%29 (accessed on 3 September 2022).

Gerhardt, C., Suhlmann, G., Ziemßen, F., Donnan, D., Warschun, M., & Kühnle, H. J., (2020). How will cultured meat and meat alternatives disrupt the agricultural and food industry? *Industrial Biotechnology, 16*(5), 262–270.

Ghaani, M., Cozzolino, C. A., Castelli, G., & Farris, S., (2016). An overview of the intelligent packaging technologies in the food sector. *Trends in Food Science & Technology, 51*, 1–11.

Gounden, C., Irvine, J. M., & Wood, R. J., (2015). Promoting food security through improved analytics. *Procedia Engineering, 107*, 335, 336.

Graiver, D. A., Topliff, C. L., Kelling, C. L., & Bartelt-Hunt, S. L., (2009). Survival of the avian influenza virus (H_6N_2) after land disposal. *Environmental Science & Technology, 43*(11), 4063–4067.

Gunders, D., (2012). Wasted: How America is losing up to 40% of its food from farm to fork to landfill. *Natural Resources Defense Council, 26*, 1–26.

Gustavsson, J., Cederberg, C., Sonesson, U., Van, O. R., & Meybeck, A., (2011). *Global Food Losses and Food Waste: Extent, Causes, and Prevention Food and Agriculture Organization of the United Nations*. Rome: Italy.

Hazards, E. P. O. B., (2015). Scientific Opinion on a continuous multiple-step catalytic hydro-treatment for the processing of rendered animal fat (Category 1). *EFSA Journal, 13*(11), 4307.

Helland, J., & Sørbø, G. M., (2014). *Food Securities and Social Conflict*. CMI Report.

Hennekinne, J. A., De Buyser, M. L., & Dragacci, S., (2012). Staphylococcus aureus and its food poisoning toxins: characterization and outbreak investigation. *FEMS Microbiology Reviews, 36*(4), 815–836.

Irani, Z., Sharif, A. M., Lee, H., Aktas, E., Topaloğlu, Z., Van't, W. T., & Huda, S., (2018). Managing food security through food waste and loss: Small data to big data. *Computers & Operations Research, 98*, 367–383.

Johnson, L. K., Dunning, R. D., Bloom, J. D., Gunter, C. C., Boyette, M. D., & Creamer, N. G., (2018). Estimating on-farm food loss at the field level: A methodology and applied case study on a North Carolina farm. *Resources, Conservation, and Recycling, 137*, 243–250.

Johnstone, S., & Mazo, J., (2011). Global warming and the Arab spring. *Survival, 53*(2), 11–17.

Jurgilevich, A., Birge, T., Kentala-Lehtonen, J., Korhonen-Kurki, K., Pietikäinen, J., Saikku, L., & Schösler, H., (2016). Transition towards circular economy in the food system. *Sustainability, 8*(1), 69.

Lagerberg, F. C., Vågsholm, I., & Birgersson, A., (2011). *Från förlust till vinst: Såhär minskar vi matsvinnet i butik.*

Lebersorger, S., & Schneider, F., (2014). Food loss rates at the food retail, influencing factors and reasons as a basis for waste prevention measures. *Waste Management, 34*(11), 1911–1919.

Leibler, J., Dalton, K., Pekosz, A., Gray, G., & Silbergeld, E., (2017). Epizootics in industrial livestock production: Preventable gaps in biosecurity and biocontainment. *Zoonoses and Public Health, 64*(2), 137–145.

Lipinski, B., Hanson, C., Waite, R., Searchinger, T., & Lomax, J. (2013). *Reducing Food Loss and Waste.* World Research Institute, 1–40.

Losses, F. G. F., & Waste, F., (2011). *Extent, Causes, and Prevention.* Rome: Food and Agriculture Organization of the United Nations.

Lovins, A. B., (1990). The negawatt revolution. *Across the Board, 27*(9), 18–23.

Manning, L., & Soon, J. M., (2016). Food safety, food fraud, and food defense: A fast-evolving literature. *Journal of Food Science, 81*(4), R823–R834.

Martínez, Z., N., Menacho, P., Z., & Pachón-Ariza, F., (2014). Food loss in a hungry world, a problem? *Agronomía Colombiana, 32*(2), 283–293.

McDermott, J. J., Staal, S. J., Freeman, H., Herrero, M., & Van De, S. J., (2010). Sustaining intensification of smallholder livestock systems in the tropics. *Livestock Science, 130*(1–3), 95–109.

Moffitt, C. M., & Cajas-Cano, L., (2014). Blue growth: The 2014 FAO state of world fisheries and aquaculture. *Fisheries, 39*(11), 552–553.

Monsees, H., Kloas, W., & Wuertz, S., (2017). Decoupled systems on trial: Eliminating bottlenecks to improve aquaponic processes. *PloS One, 12*(9), e0183056.

Mourad, M., (2016). Recycling, recovering, and preventing "food waste": Competing solutions for food systems sustainability in the United States and France. *Journal of Cleaner Production, 126*, 461–477.

Nychas, G. J. E., Panagou, E. Z., & Mohareb, F., (2016). Novel approaches for food safety management and communication. *Current Opinion in Food Science, 12*, 13–20.

Oivanen, L., Mikkonen, T., & Sukura, A., (2000). An outbreak of trichinellosis in farmed wild boarin Finland. *APMIS, 108*(12), 814–818.

Parfitt, J., Barthel, M., & Macnaughton, S., (2010). Food waste within food supply chains: Quantification and potential for change to 2050. *Philosophical Transactions of the Royal Society B: Biological Sciences, 365*(1554), 3065–3081.

Poyatos-Racionero, E., Ros-Lis, J. V., Vivancos, J. L., & Martínez-Máñez, R., (2018). Recent advances on intelligent packaging as tools to reduce food waste. *Journal of Cleaner Production, 172*, 3398–3409.

Rockström, J., Williams, J., Daily, G., Noble, A., Matthews, N., Gordon, L., & Steduto, P., (2017). Sustainable intensification of agriculture for human prosperity and global sustainability. *AMBIO, 46*(1), 4–17.

Rocourt, J., BenEmbarek, P., Toyofuku, H., & Schlundt, J., (2003). Quantitative risk assessment of *Listeria monocytogenes* in ready-to-eat foods: The FAO/WHO approach. *FEMS Immunology & Medical Microbiology, 35*(3), 263–267.

Ropodi, A., Panagou, E., & Nychas, G. J., (2016). Data mining derived from food analyses using non-invasive/non-destructive analytical techniques; determination of food authenticity, quality & safety in tandem with computer science disciplines. *Trends in Food Science & Technology, 50*, 11–25.

Rosengren, Å., (2017). *Tillväxt av bakterier under avsvalning, förvaring och upptining. Livsmedelsverkets Rapport, 2.*

Salemdeeb, R., Zu Ermgassen, E. K., Kim, M. H., Balmford, A., & Al-Tabbaa, A., (2017). Environmental and health impacts of using food waste as animal feed: A comparative analysis of food waste management options. *Journal of Cleaner Production, 140*, 871–880.

Scherhaufer, S., Moates, G., Hartikainen, H., Waldron, K., & Obersteiner, G., (2018). Environmental impacts of food waste in Europe. *Waste Management, 77*, 98–113.

Schneider, F., (2013). The evolution of food donation with respect to waste prevention. *Waste Management, 33*(3), 755–763.

Scialabba, N., (2015). *Food Wastage Footprint & Climate Change* (pp. 15–19). UN FAO.

Segrè, A., & Gaiani, S., (2012). *Transforming Food Waste into a Resource*. Royal Society of Chemistry.

Sundström, J. F., Albihn, A., Boqvist, S., Ljungvall, K., Marstorp, H., Martiin, C., & Magnusson, U., (2014). Future threats to agricultural food production posed by environmental degradation, climate change, and animal and plant diseases: A risk analysis in three economic and climate settings. *Food Security, 6*(2), 201–215.

Tabatabaei, M., Karimi, K., Kumar, R., & Horváth, I. S., (2015). *Renewable Energy and Alternative Fuel Technologies*. Hindawi.

Unaegbu, E. U., & Baker, K., (2019). Assessing the potential for energy from waste plants to tackle energy poverty and earn carbon credits for Nigeria. *Int. J. Energy Policy Manag., 4*, 8–16.

Vågsholm, I., Arzoomand, N. S., & Boqvist, S., (2020). Food security, safety, and sustainability—getting the trade-offs right. *Frontiers in Sustainable Food Systems, 4*, 16.

Von, B. J., (2010). Food insecurity, hunger, and malnutrition: Necessary policy and technology changes. *New Biotechnology, 27*(5), 449–452.

Wadhwa, D. R., & Smith, M. A., (2017). Pregnancy-related listeriosis. *Birth Defects Research, 109*(5), 324–335.

Wooldridge, M., Hartnett, E., Cox, A., & Seaman, M., (2006). Quantitative risk assessment case study: Smuggled meats as disease vectors. *Revue Scientifique Et Technique-Office International Des Epizooties, 25*(1), 105.

Ye, Y., Barange, M., Beveridge, M., Garibaldi, L., Gutierrez, N., Anganuzzi, A., & Taconet, M., (2017). FAO's statistic data and sustainability of fisheries and aquaculture: Comments on Pauly and Zeller 2017. *Marine Policy, 81*, 401–405.

CHAPTER 6

Impact of Food and Agricultural Wastes on the Environment: Management Strategies and Regulations to Curb Wastes

YASMEENA JAN,[1] MUNEEB MALIK,[1] AFROZUL HAQ,[1] BIBHU PRASAD PANDA,[2] MIFFTHA YASEEN,[1] ENTESAR HANAN,[1] ISHFAQ YASEEN,[3] and ASIF RAFIQ[4]

[1] *Department of Food Technology, School of Interdisciplinary Sciences and Technology, Jamia Hamdard, New Delhi, India, E-mail: yasmeenajan070@gmail.com (Y. Jan)*

[2] *Department of Pharmacognosy and Phytochemistry, School of Pharmaceutical Education and Research, Jamia Hamdard, New Delhi, India*

[3] *Department of Management, Islamia College, University of Kashmir, Jammu and Kashmir, India*

[4] *College of Temperate Sericulture, Sher-e-Kashmir University of Agricultural Sciences and Technology, Shalimar, Srinagar, Jammu and Kashmir, India*

ABSTRACT

An enormous food and agricultural waste are produced annually around the globe. Waste arising from food and agricultural sectors has the potential of vast application. India is one of the largest agricultural-driven

---

Integrated Waste Management Approaches for Food and Agricultural Byproducts. Tawheed Amin, PhD, Omar Bashir, Shakeel Ahmad Bhat & Muneeb Ahmad Malik (Eds.)

© 2023 Apple Academic Press, Inc. Co-published with CRC Press (Taylor & Francis)

countries with diverse geographical regions with varying climates around the year. Harvesting of agricultural produce is either followed by the processing of the fresh produce or storage for later consumption. All these activities result in a vast generation of waste in terms of crop residue or food waste. In countries like India, the generation of consumable and non-consumable agricultural wastes throughout the supply chain, from farm to the consumer, is very tremendous. Various studies carried out by the National Engineering and Environmental Research Institute (NEERI), Nagpur; Central Pollution Control Board (CPCB), New Delhi; Central Institute of Plastics Engineering and Technology (CIPET), Chennai; and Federation of Indian Chambers of Commerce and Industry (FICCI), New Delhi, have suggested an increasing trend of the waste generation by the urban sector including the household waste and the other solid wastes. Globally, around 1.3 billion tons of food, which count to about 33% of the edible food, is being discarded during the entire food supply chain (FSC). Incineration of agricultural residue, dumping, and burial of the solid and the other food industrial wastes has resulted in large-scale land, water, and air contamination. Serious environmental degradation and alarming situation has forced the governments to take concrete measures in order to overcome/reduce the wastage and enhance the feasible management practices for the conversion of non-edible/waste material into functional products and energy. Though, efficient execution of these methods needs strong policies and strategies in addition to other socio-economic factors. This chapter addresses the essential technological aspects, possible solutions, and sound policy concerns to accomplishing long term way out of waste management (WM) and to minimize the negative impact of waste on the environment.

6.1 INTRODUCTION

India ranks second in the world's most largely populated countries with a population of more than 100,000 crores, which accounts to nearly about 18% of the world's population. India has five large megacities Mumbai, Delhi, Chennai, Kolkata, Bengaluru, 53 small cities, 640 districts, and 7,935 towns with urban population (31.2%) counting up to 377 million (Census, 2011). To tackle the food demands of this rising population, there has been an increase in agricultural activities and hence food production. India being diverse in its climate and culture, the population living in these zones of varied climate has diverse eating patterns and also waste generation. A considerable amount of waste is generated from the agricultural activities like farming residues, industries, or households (Bhuvaneshwari et al., 2019; Joshi et al., 2016). The various public bodies such as municipalities mostly govern and manage the waste generated as municipal waste and thus, data of waste generated is assembled, investigated, and documented in the domain of public. The farming waste is primarily handled by the agricultural landowners who principally fall in the private domain with less available data. The increasing demand of foodstuffs both in developed and underdeveloped nations has led to immense raise in ration production throughout the globe. Because of this, the agriculture-based practices also correspond to very beneficial as well as profitable business sources. Huge areas of the inhospitable surroundings (land) are being converted into arable lands mainly because of progress in water managing systems, new agro techniques, and extensive agrochemical operations. These operations have resulted in an increase in worldwide environmental contamination and created greater than before difficulty in dumping of the waste generated. The massive amount of agricultural waste finally paved the way to large scale pollution of air, water, and land. The patterns of the farming activities employed and the waste generated mainly reflects the environmental and civilizing aspects of the country. Despite of these environmental issues, the various national organizations are constantly putting forth different policies and many possible alternatives to manage this waste, such as transforming waste into recyclable as well as reusable resources. As per the United Nations, the entire residue generated as a result of a range of farming activities is termed as the agricultural wastes. It comprises of manure and other wastes from poultry, farmhouse, slaughterhouse, harvest waste, chemical pesticides residues, and chemical fertilizer runoff, salt, silt drained from fresh produce (Korres, 2020). As per world energy council, farming waste also encompasses of disfigured food wastes (non-edible foods) (Hoornweg and Bhada-Tata, 2012). The crop residue comprises of both the field residue as well the residue that is being produced after the processing of the crop into the useful products. Stalks stems, pods leaves, straw, rice straw are some of the examples of field residues. The process residues mainly comprise of residues such as sugarcane bagasse and molasses brewery wastewater from beverage industries (United, 1997; Hoornweg, Bhada-Tata, 2012; Agamuthu, 2009; Obi et al., 2016). As per FAO, 1.3 billion tons of food is wasted each year globally that directly impacts to food shortages, water dearth, loss of biodiversity and increased greenhouse gas emissions. Reducing food and agricultural waste at primary level, producing biogas through anaerobic digestion (AD), composting or transforming to animal feed are some of the most effective ways of WM.

6.2 SCENARIO OF AGRICULTURAL AND FOOD WASTES IN INDIA

In India, agricultural sector is the major contributor to the overall economy. According to National Horticulture Database (2014), India is the largest producer of fresh produce after China. For the year 2014, India produced about 88.97 and 162.98 million metric tons of (MMT) fruits and vegetables (F&V), respectively, which comprise nearly 12.6% and 14% of the whole global production of F&V, respectively (Balaji et al., 2016). India is also among the leading producers of numerous crops such as rice, wheat, cotton, pulses, and sugarcane. Regardless of this huge quantity of produce, the quantity of exportation to other countries is only 1–2%. More than 18% of fresh produce generated from farm fields is discarded from the start of the post-harvest supply chain till their final stage. Moreover, large areas of the forest land are being used to produce food grains, causing deforestation. It also resulted in indefensible farming activities, and unnecessary extraction of water (groundwater) to meet the growing need of food. Near about 300 million barrels of fuel are being utilized to make food, which eventually gets discarded as waste (Vaidyanathan, 2018). Majorly, the non-efficient management and poor integrated approach of supply chains are the main factors responsible for this much of wastage. Reports from the Associated Chambers of Commerce and Industry of India (ASSOCHAM) in year 2013 stated that India acquires huge post-farming losses every year primarily due to the shortage of food industries and proper storage facilities (ASSO-CHAM, 2013). Round the globe, there is enough production of the foodstuffs to meet up the basic food requirement of twice the world's populace, however due to wastage, food scarcity is seen among billions of populations who are underfed, and half-starved. Near about 40% of the foodstuff produced in India is discarded as per report of the United

Nations Development Program and up to 21 million tons (MT) of wheat are discarded in India and around 50% of all foodstuff throughout the globe meet up the similar fate and by no means reaches the poor people. As per reports of the agriculture ministry, food produced of worth INR 50,000 crores is being discarded as waste each year in the country. The toll of the starving populace in India has risen by 65 million, which is larger than the population of France. The study conducted by the Bhook (an organization operational towards the reduction of hunger in 2013) observed that a large portion of the Indian population (20 crores) sleep hungry and about 7 million children passed away in 2012 because of starvation/malnutrition.

Since the food and agricultural waste go analogous with each other, it can easily be evaluated that there is also wastage of the various farming produce expressed as the crop residues. As per the reports of the Indian Ministry of New and Renewable Energy (MNRE), the agricultural sector in India yields near about 500 million tons (Mt) of harvest left over each year. In the same study, it showed that a greater part of these crop leftovers are utilized as fodder, fuel for other household and built-up functions. Though, there is still an additional of 140 Mt out of which 92 Mt is burned each year. Figure 6.1 shows the farming waste yielded by specific Asian countries in Mt/year. It is also attention-grabbing that the portion burnt as farming residues in India is much larger than the overall production of farming residues in comparison with other nations in the area (Bhuvaneshwari et al., 2019).

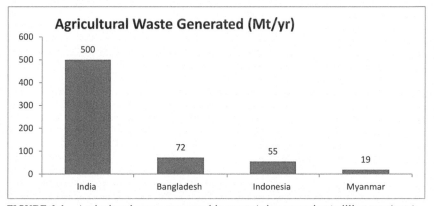

FIGURE 6.1 Agricultural waste generated in some Asian countries (million tons/year).

6.3 CLASSIFICATION OF AGRICULTURE AND FOOD WASTES (AFW)

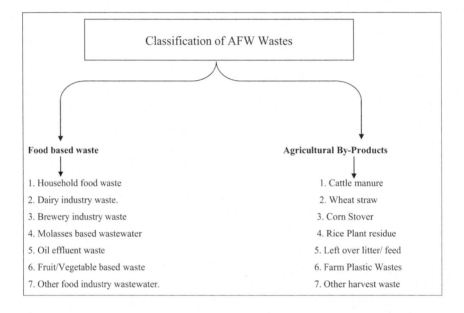

Agriculture and food wastes are classified into two types: food-based waste and agricultural by-products.

- Food-based waste:
  - Household food waste;
  - Dairy industry waste;
  - Brewery industry waste;
  - Molasses-based wastewater;
  - Oil effluent waste;
  - Fruit/vegetable-based waste;
  - Other food industry wastewater.
- Agricultural by-products:
  - Cattle manure;
  - Wheat straw;
  - Corn stover;
  - Rice plant residue;

- Leftover litter/feed;
- Farm plastic wastes;
- Other harvest waste.

6.4 FOOD WASTE

6.4.1 SOURCES OF WASTE

The FAO describes food waste as "wholesome edible material intended for human consumption, arising at any point in the FSC that is instead discarded, lost or degraded" (Gustavsson et al., 2011). These can also be defined as the amounts of safe-to-eat substances discarded or vanished in the FSC at various pre and post-harvest stages, during produce handling, processing stages and finally at the end-stage. (Balaji et al., 2016). The various food wastes include leftovers of fresh fruits and vegetables from canning industries, cereal grains, bakeries (stale bread, cookies, etc.), paper napkins, coffee filter paper, eggshells or egg trays, and other packaging materials (newspaper) generated during various operations which can be composted. The main factors responsible in the contribution of the food waste are the sequence of the processing of the fresh produce, improper storage of the fresh produce, cross-contamination, poor packaging, handling during the transportation (bruising, crushing, etc.) and overall, the greater inventory owing to pitiable forecasts (Papargyropoulou et al., 2014). The by-products which don't have been recycled or used for other functions/purposes are usually being generated by multiple food operations carried out in the various food-processing units. As such, various food units contribute to the production of the greater volumes of waste and also other solids consequential from the production, training, and utilization of food. The waste of food being rich in moisture makes it more feasible for making the compost with the help of bulking agent (high ratio of Carbon: Nitrogen example sawdust and yard waste) which helps in the absorption of the moisture and adding the structure to the compost (Shilev et al., 2007). These enormous amounts of wastes cause rising dumping and likely intense pollution difficulties and correspond to thrashing of important biomass and nutrients (Litchfield, 1987).

6.4.2 CAUSES OF FOOD WASTE

While FL occurs mostly at the time of production due to lack of skills and proper infrastructure, poor practices or natural calamities, food waste happens when food that is palatable is deliberately discarded by consumers when they do not carefully plan their meals, and stock food till it deteriorates or expiry date is over. Food waste also occurs due to glut in markets during peak seasons. At times, over-preparation of food in hotels, restaurants, and the foodservice industry results in food wastage and sometimes due to substandard quality and esthetic look, a lot of food is rejected by the retailers. FAO in a 2013 report analyzed the impact of global food wastage on the environment. It was estimated that near about 1/3rd of the entire world food production is wasted or does not find its way to our tables.

6.4.3 WAYS TO MINIMIZE FOOD WASTE

- There should be a balance between production and demand. Excess production finally ends up getting wasted. In case of over-production, measures should be taken to redistribute or to divert the food to people in need.

- Additional effort to develop better facilities (harvesting, processing, storing, and distributing) to minimize food waste are needed.

- Effective measures must be taken at every single stage of the process to stop food waste from growers/food processors to super-stores or individual consumers.

- Identify critical points of waste generation and taking measures to tackle it.

- Consumers should always plan meal plan in accordance with their need to prevent food wastage.

- In case, the food in anyhow not fit for humans, can still be used as a livestock feed, or for producing commercial feed. If the food cannot be reused at all, then instead of sending it to the landfills where it continues to rot, it should be tried for recycling in a responsible manner. By adopting home composting on an average home can avert a huge amount of food waste in a year from disposing to waste facilities.

- Inadequacies in the FSC need to be focused. It could contribute positively and can help to reduce loss of foodstuffs from field stage till end of consumption.

6.5 AGRICULTURAL WASTE

Waste substances arising from different farming practices (straw, husk, tree pruning), from agro-industries or from livestock farming are classified as agricultural wastes. India, which has around 17% of the total world population and agricultural environment, produces huge quantity of foodstuffs (Jain et al., 2014). As per reports from the Directorate of Economics and Statistics, 2012–2013, there was production of 361 Mt of sugarcane, 94 Mt of wheat and 105 Mt of rice in India (Figure 6.2). Of the variety of crops grown, bulk crop residues come from rice, wheat, and sugarcane which are dealt by burning. These harvest produces have hefty profits in returns on investing them, thus creating a difficult situation for farm persons to yield different produce, with less crop residues.

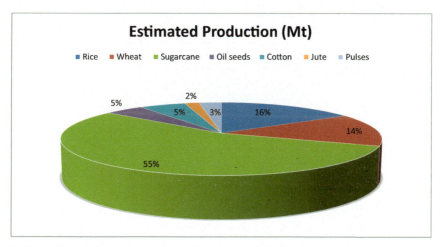

FIGURE 6.2 Estimated production of major crops in India.

National policy for management of crop residues (NPMCR) endows with the information of the harvest residues of the states. As per the report of NPMCR, the production of crop residues is maximum in Uttar Pradesh

(60 Mt), then Punjab (51 Mt) and Maharashtra (46 Mt), with a total sum of 500 Mt every year out of which, 92 Mt is destroyed by fire (Bhuvaneshwariv et al., 2019). Major crops (Rice and wheat) contribute to almost 70% of the crop residues (Figure 6.3). Apart from overall farm residue generated, the excess crop residue implies to the remains which are left after consuming the residue over a number of activities such as fuel for industries, households, etc. A portion of this discarded crop residue is fired up and left portion after burning is kept as such in the fields.

FIGURE 6.3 Crop residues produced by major crops.

6.6 IMPACT OF FOOD AND AGRICULTURE WASTE ON ENVIRONMENT

During the last century, the global populace has increased several folds and to meet the demands of more food for such a huge population, there has been increased food production and overall agricultural activities. The waste generated from food and agricultural activities has a very drastic impact on the environment because of the excess emissions of the various greenhouse gases (GHGs) and other hazardous emissions causing social,

Impact of Food and Agricultural Wastes on the Environment 175

economic, and ecological imbalances. Following are the summarized effects/impact of various food and agricultural wastes on the environment:

- As per estimates of the United Nations, one in nine persons in the world do not have access to adequate food to live a healthy life. Near about one-third of the food produced throughout the world is wasted or lost due to one or other reason. It is reported that more people die from starvation each day than tuberculosis, AIDS, and malaria altogether. Food wastage or FL not only cause massive economical losses, but also has severe negative impact on environment.
- During the upstream process (production, handling, and storage), more than 50% of the food wastage takes place and the remaining occurs during downstream (processing distribution and consumption) phase.
- Greater the food wastage at downstream stage, more will be its impact on environment, because the energy and resources used in processing, transporting, storage, and cooking are also lost.
- As per FAO report 2013, middle-and higher-income nations showed higher FL and waste during the downstream stage due to lack of proper harvest techniques and infrastructure in developing nations and wastage or FL happens during upstream stage.
- An enormous volume of freshwater and groundwater gets wasted, which is used just to produce food that is not eaten. Near about 50,000 liters of water get wasted by throwing out one kilogram of beef that were used to produce it. In the same way, pouring down the drain a glass of milk, nearly 1,000 liters of water are wasted.
- Food waste that ends up in landfills produces a large amount of greenhouse gasses (methane and carbon dioxide) which are responsible for global warming and climate change.
- Around 1/3rd of the world's total agricultural land area (1.4 billion hectares) is used to grow food that is wasted. Gallons of oil and tons of chemical fertilizers and pesticides are also wasted each year to produce food that is not eaten. The enormous amount of agricultural activities led to rise in environmental pollution and waste production.
- Furthermore, food waste has negative impacts on biodiversity due to activities like mono-cropping and converting wildlands into agricultural areas.

- Outsized broadened areas of inhospitable surroundings have been changed to arable areas due to improvements in water managing schemes, up-to-date and advanced agricultural equipments and huge amounts of chemicals used for agricultural fields. These events have ended up, resulting in augmented difficulty in the dumping of agro-residue, environmental pollution globally and hence the ecologically unbalanced environment.
- India ranks second in the production of major crops (rice and wheat) in the world. These two crops typically generate huge amount of residues, and their indefensible managing practices significantly add up more burden to the environment (Bhuvaneshwari et al., 2019).
- Crop residue burning has become a major ecological as well as environmental misfortune, causing health issues and a contributor to global warming.
- The worst impact by the burning of these residual wastes (both from food and agricultural) is on the quality of the air. In India, the allowable levels of $PM_{2.5}$ in the air is 40 $\mu g/m^3$ (National Ambient Air Quality Standard) and 10 $\mu g/m^3$ (WHO standard). The capital of the country, Delhi evinced a level of 98 $\mu g/m^3$, which is double the set values and 10 times greater than WHO set values.
- A number of air contaminants like CO, CO_2, NH_3, SOX, NOX, volatile organic compounds (VOCs), non-methane hydrocarbon, and particulate matter (PM) considerably amplify as a result of burning of crop residues (Mittal et al., 2009; Zhang et al., 2011). These hazardous gases chiefly account for the degradation of the essential soil nutrients, including organic carbon or nitrogen, thus decreasing soil fertility (Jain et al., 2014).
- The most important unfavorable effects of burning of crop residue includes the emission of GHGs that adds to the global increase in temperature, greater than before smog and PM levels that raises the human health issues, hammering of biodiversity of farming lands, and worsening of soil productiveness (Lohan et al., 2018).
- Crop burning elevates the PM in the air and rises up in considerably to seasonal variations. One of the major factors responsible for the climate change is the emission of fine particles of primary and secondary carbon, which also alters the light absorption (Jiang et al., 2019; Washenfelder et al., 2015).

Impact of Food and Agricultural Wastes on the Environment 177

- The heavy haze formed as a result of the smoke generated from the burning of agricultural residues in north Indian states with the beginning of winters causes hazardous pollution and poses a great threat to human health.
- Because of the emission of the gases and the generation of the excess heat as a result of burning of the residual of crops, there is depletion in the ozone layer by the release of freons and halons caused due to the incineration of the biogas in the various plants (Whiting et al., 2014).
- Agricultural wastes are usually rich in nutrients such as phosphorus, nitrogen, ammonia, and nitrates, that run straight into main water bodies ultimately leading to eutrophication in the portable water sources. This in turn leads to depletion of oxygen and suffering of aqua life because of suffocation and also affect the photosynthesis rate in plants.
- Insect repellent or chemical fertilizers, from the farming lands reach to water bodies with runaway rainwater or other means. The residence time of these non-biodegradable chemicals is longer in water bodies and hence results in the biomagnification in humans via the various food chains.

6.7 METHODS OF TREATING WASTES

The various methods which can be designed for the treating of both the agricultural and food wastes (AFWs) are listed in Figure 6.4.

6.7.1 COMPOSTING

Composting can be described as the method in which there is decaying of organic material in the presence of a favorable environment for debris or leftover eating microorganisms to flourish (United States Department of Agriculture, USDA). The end product of this extreme decaying procedure is nutrient-rich soil that can facilitate the growth of crops, plant life and vegetation to cultivate (https://www.livescience.com/63559-composting. htm). During composting, the natural materials (plants, crop residues) undergo various changes which are generally categorized into two distinct stages:

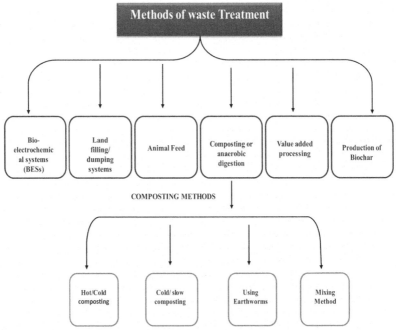

FIGURE 6.4 Schematic diagram of agriculture and food waste treatment.

1. **Degradation:** The initial stage of the composting method begins with the deterioration of the mainly simple decaying organics (sugars, organic acids, amino acids). The process is being aided by aerobic microorganisms with the consumption of oxygen and emission of CO_2 and energy. This stage is a high temperature phase (thermophilic phase). The duration of the phase lasts from weeks to several months and relies on the state of the substrate distinctiveness. The forceful airing or addition of the composting components is compulsory to lower the temperature of the substrate. In addition to this, there should be proper maintaining of the oxygen to the substrate. The prime factors like increased temperature, humidity, suitable pH, are responsible for thriving bacteria during the decaying process.

2. **Maturation:** The final stage that is the maturation is further differentiated with breakdown of the decomposed substance. The development of small and fine particles is aided by number of invertebrates (earthworms, ticks, and centipede). When the

organic matter attains maturity, followed by humification, the humus development is robustly linked to elemental and organic translation of animal or plant waste by microbial or fungal action on lignin or other organic substances (Figure 6.5).

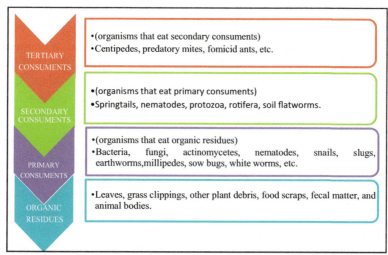

FIGURE 6.5 Series of consumers of food waste in composting process.

6.7.1.1 COMPOSTING PARAMETERS

1. **Temperature:** Organic matter is broken down by microorganisms during composting to CO_2, water, and heat. Under ideal settings, the temperature of composting varies during the process and its stages can be differentiated as:

   i. The moderate temperature period that lasts for a couple of days;
   ii. The high temperature period that lasts from a few days to several months; and
   iii. The cooling and maturation period that lasts for several months.

   Diverse microbial groups predominate during the composting stages. Mesophilic microorganisms carry out decomposition during initial stage. Biodegradable compounds are swiftly broken down and compost temperature increases rapidly during this stage as heat is produced by the action of microorganisms. As temperature reaches 40°C, these mesophiles become less active and are swapped by the thermophiles. When temperature elevates above 55°C, most of plant and human pathogens do not survive. Further increase in temperature by a few degrees kills weed seeds and other parasites. Temperature above 65°C may kill many useful microbes, as a result rate of decomposition will slow down. Thus, it is important to maintain temperature below this point by aeration and mixing. Elevated temperatures during the thermophilic stage speeds up the breakdown of plants key structural molecules such as complex carbohydrates, fats, proteins, etc. Upon depletion of these molecules, compost temperature begins to decrease gradually, and once again mesophiles come into action for the last curing/maturation of the residual organic matter.

2. **Enzymes:** These play a major role in the organic matter transformation. These enzymes are formed by active microorganism and are categorized as intra and extracellular enzymes. Bigger molecules are usually degraded by the action of extracellular enzymes. With the destruction of the larger molecules in the outer environment, smaller molecules via different mechanisms enter the microbial cell and are acted upon by intracellular enzymes.

3. **C:N Ratio:** Carbon and nitrogen are the most vital elements that must be present during the microbial decomposition. Carbon acts as an energy source and represents half of the building material of microbial cell biomass. Apart from carbon, nitrogen is an important constituent of the proteins, enzymes, coenzymes, and nucleic acids essential for the growth and functioning of cells. It is important to know the C:N ratio of the compost ingredients so that they are blended in appropriate amounts to obtain this optimum ratio of the compost. For each compost ingredient, the optimum carbon to nitrogen ratio is necessary. Generally, C:N ratio of around 30:1 is considered to be optimum for composting. For the compost microbial activity, apart from carbon and nitrogen, other elements such as calcium, sulfur, and phosphorus are also important.

4. **Oxygen and pH:** For successful composting, oxygen is another crucial component as it is an aerobic process. The oxygen content of 10% is considered optimal, while a pH optimal for compost microorganisms is between 5.5 and 8.5. When the organic matter

is digested by the bacteria and fungi, organic acids are released. As more acid accumulates, this results in a decrease of pH to 4.5, which limits the microbial activity. Introduction of air in such cases is generally adequate to bring the pH of the compost to satisfactory levels.

5. **Humidity:** One of the most fundamental factors for the survivability of compost's active microorganisms is water, because:

   i. For the soluble substrates and extracellular enzymes, it serves as a transport environment;
   ii. It is crucial for nutrient exchange through cell membranes;
   iii. It serves as an environment for chemical reactions to take place.

   For the compost, optimal humidity is 50–60% and an increase of more than 65%, can establish anaerobic conditions, while reduction below 40% lowers the compost's biological activity. During the process of composting, humidity levels decrease naturally and may reach to levels hazardous for the biological activity. At that point, it becomes important to supplement water to raise humidity levels.

6. **Particle Size:** Composting material, total volume, and moisture content of composted material directly influence the particle size. Higher the compost compactness, finer, and wetter are the composted particles. 25–75 mm particle size is considered optimal. Addition of more inert materials such as sawdust or bark, is necessary when the composted organic matter is of smaller size.

7. **Total Free Space:** The relation between the volume of free space and the volume of biomass is termed as total free space and is expressed as percentage. As a part of total free space, volume occupied by the air is termed as free air space, which is an important parameter that influences the oxygen retaining capacity of the compost. Normally, the value between 35% and 50% is considered optimal for total free space (Keener et al., 2000).

8. **Stability:** Microbial stability directly influences the rate of degradation of organic matter, lack of optimal conditions retards the microbial growth processes. The processes of organic matter degradation are influenced negatively by the presence of higher concentrations of heavy metals, which lead to lethal effects,

resulting in restriction of microbial activity (Chiumenti and Chiumenti, 2002; Scaglia et al., 2000).

6.7.1.2 METHODS OF COMPOSTING

1. **Slow Composting:** Slow (cold) composting is suitable when composting material is relatively rich in carbon than nitrogen. The benefit of slow composting lies in its easy handling and management. The drawbacks include a slow rate of decomposition and pest infestation. Moreover, plant pathogens or seeds of weed do not get destroyed in slow composting process. Such kind of composting could be carried out in cold bins, sheet trend or in heaps. Organic matter builds up throughout the yard, in washed out areas, under trees, or in the space near the garden, etc. The material will decompose, in a year or two supplying valuable organic matter to the soil.

2. **Fasts Composting:** Fast /hot composting yields compost at a much faster rate and provides the greatest control over plant pathogens and weed seed. Fast composting is a rigorous process and needs various elements, such as; at least of 1 m³ of material to start the pile, suitable C:N ratio (obtained by blend of greens and browns), particle size of lesser than 2 to 3 cm, appropriate moisture content and regular turning to provide adequate aeration to succeed.

3. **Mixing Method:** To speed up the composting process, the mixing method is the most appropriate. Organic matter with different moisture and C:N ratio, is blended and put to the compost system. Mixing avoids compact layers creation that may limit the flow of oxygen and water through the mass. Mixture is added to a compost system in batches. For even distribution of moisture, each batch needs to be watered. Fresh materials can also be added to an active composting pile during mixing or turning. Fresh material addition will supplement the existing base. If a large amount of material rich in carbon are used, it will dry out the pile, reducing the pile temperature and ultimately resulting in slowing the composting process. If a large amount of material with low C: N ratio are used, it will increase nitrogen concentration, heat up the pile resulting in speeding up the composting process.

Impact of Food and Agricultural Wastes on the Environment 183

4. **Using Earthworms:** Worm composting or vermin-composting is a process where worms are used to recycle food and agriculture scraps to amend the soil composure. Worms eat the organic waste and transform that to valuable soil. Worms are phenomenal and unparalleled decomposers, specifically African nightcrawlers and red wigglers, when table and kitchen scraps are among the composting materials they are often used.

6.7.2 BIOELECTROCHEMICAL SYSTEMS (BESS)

These are the diverse set of biologically catalyzed electrochemical systems with wide range of functions that are intended for substrate removal and energy conservation. BESs directly act on organic substrate and convert the chemical energy into electrical energy via a series of redox reactions (Mohan et al., 2014; Pant et al., 2012). The electron-donating and accepting phenomenon is linked with microbial metabolism via artificially introduced electrodes, which induces potential difference that causes bioelectrogenic activity. A number of soluble complex organic wastes and renewable biomass can be exploited as a substrate by microbial fuel cells to offer an eco-friendly process with dual benefits of waste remediation and renewable energy generation (Mohan et al., 2014). During the last few years, the capability of BESs to harness the electrons produced via electrodes has improved exponentially (Logan and Regan, 2006). For use in BESs, a good number of substrates have been explored. Food and agriculture waste are of particular interest. Amongst promising technologies developed in recent times, BESs has come out as a system with multiple applications like wastewater treatment, bioenergy generation, and synthesis of value-added products (Pant et al., 2012).

6.7.3 VALUE-ADDED PROCESSING

Another method of management of food and agricultural waste is by converting it to different eco-friendly value-added products such as bioactive compounds, single-cell proteins, phytochemicals, bio-pigments, dietary fibers, food, and feed supplements, essential oils, biofertilizers, and biofuels. For example, drying of discard and other waste like from potatoes is a very efficient and effective method of transforming waste material into a value-added product such as potato granules and flakes.

A number of new re-composition products from potato and other food wastes are produced nowadays in the food industries.

6.7.4 ANIMAL FEED

The practice of converting food and agriculture waste into animal feed is not new. Many countries have now processing and feed production facilities to convert waste from different stages of chain supply into feed for animals. Waste from potatoes and cull beans, dairy processing waste, vegetable and fruit waste, bakery waste and garbage could be processed into animal feed. Cull potatoes and other vegetable/plant wastes are an excellent source of energy that can be ensilaged. There are a number of benefits of recycling waste through animal feed such as reduced pressure on landfills, lowered carbon footprints by reducing methane emission from waste, money-saving for cattle farmers.

6.7.5 BIOCHAR PRODUCTION

As a measure to control greenhouse gasses emissions, the research community is trying continually to develop/improve ways for efficiently enhancing rates of carbon appropriation in the soil. To stabilize soil organic content, the awareness in applying biochar, charcoal, and black carbon as soil alterations has increased. To decline the greenhouse gas emissions while lessening the size of agricultural waste considerably, these techniques are viewed as the practicable options. These incompletely burnt products (pyrogenous) with a low chemical conversion rates are perfect for soil amendment (Izaurralde et al., 2001; McHenry, 2009). High resistance to chemical oxidation and residence time of biomass, are essential for the process of carbon sequestration to CO_2 or reduction to methane, which in turn reduce the CO_2 or methane discharge to the environment (Srinivasarao et al., 2013). At low temperatures and oxygen concentrations thermochemical alteration (pyrolysis) produce a carbon rich, porous, and fine-grained product known as Biochar (Amonette et al., 2009). Biochar is rich in carbon, oxygen, hydrogen, nitrogen, sulfur, and ash (Boateng et al., 2015). Extremely porous nature of the biochar aids in increased surface area and improved water retention capacity of the soil. Biochar sets off a number of biogeochemical processes and helps in nutrient retention (Maroušek et al., 2019). It also increases in pH, soil microbes and earthworm population, and decreases the excess usage of fertilizers (Gaunt and Cowie, 2009).

Impact of Food and Agricultural Wastes on the Environment

6.8 TREATMENT OF FARM PLASTICS

A number of waste materials are generated at farms, such as plastics, empty containers, chemicals, old machinery, building materials, petroleum wastes, and products of animal health care. Farm plastics are put to use in a number of ways on the farm. A new technique used in agriculture is placing plastic films on the top surface of the soil to boost heat retention. Plastics are also used for covering and storing of forages (Jacobs and Associates, 2001).

6.8.1 PROPER DISPOSAL OF FARM PLASTICS

The most suitable method for farm plastic disposal is to transform the waste into a practical by-product like toys, building materials, recreational furniture, or recycled silage wrap, etc. Burning or landfilling of farm plastics is not suggested.

6.8.2 RECYCLING OF PLASTIC WRAP

The plastic wrap is detached and contaminants like (dirt, haulage, ice, water, etc.) are removed. Plastic wraps are kept clean, dry, and stored indoors to keep it away from further contamination. Wraps are finally compacted to small sizes for easier handling, storing, and transportation.

6.8.3 RINSING OF PESTICIDE CONTAINERS

The most common type of containers used to accommodate liquid pesticides is made from plastic. Granular or powder form pesticides are held in paper bag containers. Rinsing is done to remove residual pesticides left in the containers after emptying. Rinsing is usually done instantly after use as waiting too long will cause pesticide solution to dry out, making rinsing difficult and reducing the possibility of meeting the cleaning standards. Removing pesticides prevent:

1. **Money Wastage:** A lot of money can be saved by not throwing away an empty container.

2. **Exposure:** Poison container can be hazardous to people, livestock, or wildlife.

3. **Contamination**: Soil, surface water, or groundwater can get contaminated by pesticide leftover in the container.

6.8.4   DISPOSING OF PLASTIC CONTAINERS

Returning empty plastic pesticide containers to the pesticide dealer for recycling is highly advised. In many countries, makers and dealers of pesticides have setup programs of container collection with the aim to recycle empty plastic containers. Containers must be clean, rinsed thrice with no liquid material inside. In some countries, it is mandatory for all licensed pesticide dealers to accept empty plastic containers.

6.8.5   DISPOSING OF PAPER BAG CONTAINERS

Disposing to the landfill site or energy plant is the recommended way for empty paper bag containers. Single rinsing of paper bag containers with plastic or foil lining is advised. Trace amounts of the pesticide may remain even after rinsing. Since all containers cannot be recycled so they should be crushed or punctured so that they are not used again for other purposes. An emptied pesticide container that cannot be recycled needs to be disposed off to a provincial landfill location.

6.8.6   BURNING CONTAINERS

Burning of plastic containers or any other hazardous substances on the farm is prohibited under the Environmental Protection Act – Air Quality Regulations. Burning of containers at low temperatures does not result in their destruction but in vaporization, which can drift to other areas with air movement. Also, inhaling of smoke from these fires could prove harmful.

6.8.7   BURYING CONTAINERS

The burial of emptied containers of pesticides is not recommended, even if they have been properly rinsed. Even though, appropriately cleaned containers do not pose any threat to the environment. However, their rate of decomposition is terribly slow. Plastic-cans may take many decades to completely break down.

6.9   ADVANTAGES AND DISADVANTAGES OF DIFFERENT WASTE DISPOSING METHODS (TABLE 6.1)

Impact of Food and Agricultural Wastes on the Environment 187

TABLE 6.1 Advantages and Disadvantages of Different Waste Disposing Methods

SL. No.	Methods	Advantage	Disadvantage
1.	Composting/ anaerobic digestion	• Less land area is needed for modular construction of plant.	• Destruction of pathogenic organisms is lower and less effective than in aerobic composting.
		• Results in the production of high-grade soil conditioner as well as energy recovery.	• Less organic matter containing wastes are unsuitable.
		• There is no requirement of power in contrast to aerobic composting, where pile is necessarily turned for supplying oxygen.	• Digestion efficiency decreases if waste segregation is not done.
		• All the gas produced can be put to use in an enclosed system. Thus, controlling greenhouse gas discharges.	
		• It is free from fly menace, rodents, visible pollution, bad odor and social resistance.	
2.	Landfill gas recovery	• Low price method of waste disposal	• During rainfall polluted surface run-off greatly.
		• Power can be generated by exploiting gas produced or can be used as domestic fuel.	• In the nonexistence of appropriate leachate treatment system, contamination of surface/groundwater aquifers may occur.
		• No highly skilled workers required.	• Gas recovery process inefficient, yields only 30–40% of the total gas generation.
		• Natural resources are recycled and resumed back to the soil.	• Large land area is required.
		• Low lying marshy land can be converted to valuable land.	• Balance gas including two major greenhouse gases, carbon dioxide and methane may get escaped to the atmosphere.

TABLE 6.1 *(Continued)*

SL. No.	Methods	Advantage	Disadvantage
3.	Incineration	• This is most appropriate for waste with greater calorific value and pathological wastes. • Units with high through-put and continuous feed systems can be set up. • Recovery of thermal energy for direct heating or power generation. • Comparatively odorless and noiseless. • Little area of land required. • Reduces cost of waste transportation as it can be situated within city limits. • Hygienic method of waste disposal.	• Unsuitable for waste of low calorific value or aqueous/ high moisture content wastes. • Excess of moisture and inert material affects energy recovery; supplementary fuel may be required to support sustainable combustion. • Concern for toxic material that may contaminate the environment.
4.	Pyrolysis/ gasification	• Produces gas/oil fuels, which can be put to a different usage. • In techno-economic sense, atmospheric pollution can be regulated in a much superior way as compared to incineration.	• Capital, operation, and maintenance costs are high. • Trained workers required for operation and maintenance. • For small power stations overall efficiency is low • Wastes with excessive moisture have less energy recovery.

6.10 MANAGEMENT OF AGRICULTURAL AND FOOD WASTES

As per reports of National policy for management of crop residues (NPMCR), in India, the state of Uttar Pradesh tops the list with generation of 60 Mt of crop residue wastes followed by Punjab (51 Mt) and Maharashtra (46 Mt). An overall total of 500 Mt of crop residue is generated per year in India, out of which 92 Mt is burned (Bhuvaneshwari et al., 2019). Nearly 70% of the crop residues come from rice and wheat. The total waste generated that is left after utilizing the residue for many other purposes is referred to as surplus waste. Some part of the excess waste is burned, and residual is left in the field. As per the intergovernmental panel on climate change (IPCC), of the total crop residues, more than 25% was burnt on the farm. Uttar-Pradesh state has the highest contribution to the quantity of residue burned on the farm, followed by Punjab and Haryana. The segment of crop residue burned ranges from 10–80% for paddy waste in all states. Amongst other crop residue, major contributors include rice 43%, wheat 21%, and sugarcane 19% and oilseed crops 5% (Jain et al., 2014; Sahai et al., 2011). The Ministry of Agriculture points the shortage of human labor to the increased on-farm crop residue burning (Figure 6.6). Another reason behind this is attributed to the time limit between two consecutive crop cultivations. To ensure higher economic returns, some growers resort to three crops cycle in a year with a little time gap between harvesting previous crop and sowing new one. Agriculture is responsible for 18–32% of global greenhouse gas emissions (Bellarby et al., 2008). A reduction in the food waste opens opportunities to reduce greenhouse gas emissions and diminish another negative effect of agriculture and food production on the environment. The reduction in food and agriculture waste linked emissions can be made by distinguishing between waste arising at two levels of the food system: pre-consumer waste produced during manufacturing, processing, distributing, or retailing and consumer waste produced at households.

Along with the solid wastes from the food and agricultural sector, there is also generation of the wastewaters from the various food processing units including the dairy sector, brewery wastewater, animals-based wastewater, molasses-based wastewater, vegetable-based waste, palm oil mill effluent (POME), etc. and also from the agricultural by-products that include wheat straw, cattle manure, corn stover, and soil compost. The undesirable effect of AFW on the environment makes the necessary implementation of the various management initiatives to be taken in the developing country like India (El Mekawy et al., 2015).

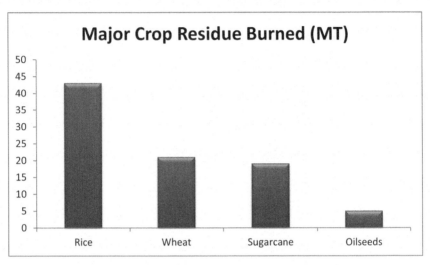

FIGURE 6.6 Data of crop residue burned as per NPMCR.

6.10.1 WASTE MANAGEMENT FUNCTIONS

AFW management consists of six basic functions these include (Figure 6.7):

1. **Production:** When waste produced is sufficiently high and become a resource concern then it necessarily requires management. Production includes the nature, volume, consistency, timing, and location of the waste produced.

2. **Collection:** The initial capture and gathering of the waste generated from the point of origin is referred as collection. The AFW plan should categorize the methods of waste collection, identify points of the collection, schedule of collection, requirements of labor, equipment, and other structural facilities, and cost of the installation of components.

3. **Storage:** It refers to the temporary holding or containment of the waste. The function of storage facility is to provide a control over

the scheduling and timing of the WM system. The WM system should recognize the storage location, period, volume, installation of storage facility and management costs.

4. **Treatment:** It includes physical, chemical, and biological treatment done to minimize the pollution or toxic potential of the waste and to increases its usage potential.

5. **Transfer:** The movement or transportation of the waste from the collection to the utilization points either as a solid, liquid, or slurry, is referred to as transfer.

6. **Utilization:** Usage of the waste for beneficial purposes and it may include recycling and reintroducing waste products into the environment.

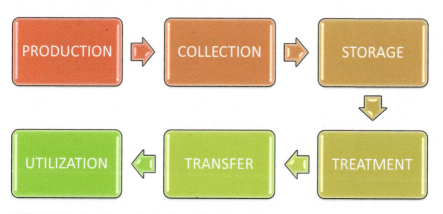

FIGURE 6.7 Functions of agricultural food waste management.

6.10.2 USE OF 3R'S APPROACH IN THE MANAGEMENT OF THE WASTES

The reduce, reuse, and recycle waste hierarchy is basically the series of steps to be taken to cut the quantity of waste generated and to enhance the overall WM programs and processes. The waste hierarchy comprises of 3R's: ***reduce, reuse, and recycle*** (Figure 6.8).

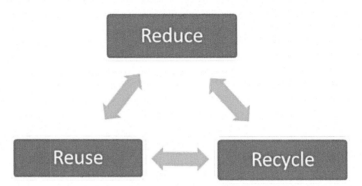

FIGURE 6.8 "3R" approach to curb waste generation.

1. **Reduce:** In waste hierarchy, it is essential to utilize the idea of reducing what is produced and consumed. The logic is simple – if waste produces is less, then there is less to reuse or recycle.

2. **Reuse:** Reusing items or repurpose them is essential in the waste hierarchy. Repurposing saves from landfills and also doesn't need any additional expenditure to dispose them.

3. **Recycling:** Recycle is the end phase of the waste hierarchy. Recycling means transforming the item again into a raw material that can be useful for mankind.

The three R's – reduce, reuse, and recycle help to minimize on the amount of waste thrown away. They conserve natural resources, landfill space and energy. In addition to this, the three R's save land and money used to dispose waste. Moreover, citing a new landfill has become more problematic and expensive due to *public opposition* and environmental protocols *(Figure 6.9).*

6.10.3 ADVANTAGES OF 3R WASTE HIERARCHY

- Minimizing the levels of greenhouse gas emissions, toxins, and pollution by significantly reducing the amount of waste thrown into the environment.
- It eliminates the practice of improper waste disposal.
- It allows managing waste in an eco-friendly manner, thus helping to protect the environment.

- The main aim of 3R's is to lower the use of new resources and to promote more effective use of already available resources.
- It enables the use of green technology, which helps to create safer and cleaner ways of waste disposal whereas minimizing the impact on the environment.
- It promotes the sustainability of energy, resources, and environment as well.
- The 3Rs boost the economy by saving energy and resources.

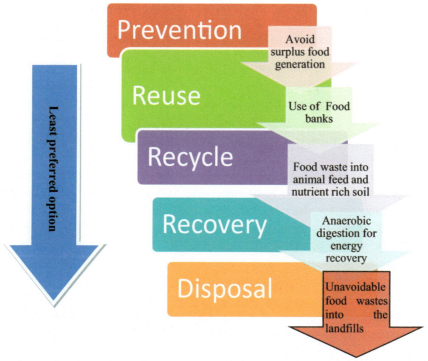

FIGURE 6.9 Promising approaches to waste management.

6.10.4 TREATMENT OF FOOD INDUSTRY WASTEWATER FOR ENERGY GENERATION

Various food industry-based wastewaters such as food waste leachate, canteen food waste, chocolate industry wastewater, yogurt waste, fermented apple juice wine lees, cereal processing wastewater, etc., have been explored as substrates for microbial fuel cells. However, the treatment efficiency and power output are dependent on the composition and nature of wastewater as well the organic load. The conversion efficiency increases with increase in the biodegradability. The energy stuck in food industry wastewater can be fetched out in the form of electricity by microbial fuel cells through significant research inputs which could benefit in closing the carbon cycle (Figure 6.10).

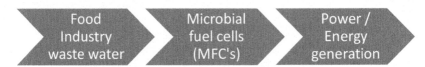

FIGURE 6.10 Energy generation from food industry wastewater.

6.11 GOVERNMENT POLICIES AND INTERVENTIONS

Food waste is a major concern in a country like India. UN estimates that about 40% of food gets wasted before reaching the consumer. However, very little attention has been paid to food waste so far, that's the reason. As per reports of 2019, India ranks 102 in global hunger index out of 117 countries. One seventh of the world's people suffer from hunger, and this is the shocking aspect of food waste (FAO, 2009). This need could be met in practical with less than 30% of food that is wasted in the world. As countries around the world develop wastage, the situation is likely to get worsened without proper policy interventions (Dorward, 2012). To minimize food waste and its emissions, there is an extensive range of policy options available. These include investment in marketing infrastructure, transportation, improved WM, better communication and integration in food chains, imparting skills to farmers for attaining the various benefits from waste, and promotion of social changes through regulations, subsidies, taxes, or consumer education (Lee and Willis, 2010). To reduce wastage and disposal emissions, developing economies will require significant infrastructure, which needs higher capital investments. However, such investments often produce secondary benefits such as opportunities of livelihood. Unnecessary demand in the food system, cause greater and more intensive use of resources for food production, which in turn promotes biodiversity loss, higher food prices, increase in conflicts between countries overland/water resources and reduces access to food for poorer people. For more healthy and productive global food system, it is significant to reduce the levels of global food waste to feed the growing world population (McLachlan and Hamann, 2011).

Food waste is a huge global issue and needs the adoption of numerous workable solutions approved by advanced nations in vindicating the effect of the food and agriculture waste. Better communication among the stakeholders, synchronization among members and more efficient food packaging could provide sustainable solutions. There is enormous potential for energy harvesting and nutrient cycling if appropriate infrastructure for waste disposal is created. The Government's involvement is necessary to improve the logistics structure to streamline the procedures for proper management and disposal of food and agriculture waste.

6.12 CONCLUSIONS

At both the local and global level, food waste is becoming an ever more important issue. Food and agriculture waste has economic implications from farmer to consumer and overall countries growth. For a developing economy like India, WM is one of the critical challenges. A systematic approach like legislative, scientific research, international collaboration, willingness of industries, and awareness among the public are the key steps to reduce the environmental problems associated with the food processing and agricultural industry, both technically and economically.

In India, plentiful food ethos has resulted in the production of huge volumes of food wastes corresponding to about 40% of the total amount of garbage produced each year. Food waste produced, is mostly either burnt or dumped in landfill sites which create various problems like putrid smell and contaminated surface and groundwater. The disposal of waste by incineration involves enormous construction costs and discharge of harmful gas and toxic ashes which is problematic. Food waste compost could act as a substitute to chemical fertilizer to improve soil quality. The economic and ecological problems caused by the waste from food processing industry can be resolved by bio-valorization of food waste to energy which has emerged as a viable alternative. The efficiency and economic trends of the system need to be further improved by integrating various industrial processes for the production of value-added products.

Agricultural wastes, if managed correctly through the application of agricultural waste management systems (AWMSs), can be transformed into beneficial products for human and agricultural usage. The solid waste in India makes up 40–85% of the waste, an important contribution could be made towards waste recovery by composting this fraction and transformation into valuable materials. Distinct collection and composting of market waste can pay positively toward environment and reduce the difficulties linked to the collection and transportation of food waste. Food and agriculture waste prevention and management require holistic and sustainable solutions and, foremost, a fundamental re-think of the existing practices, policies, and systems in place for addressing the waste challenges and to deliver considerable social, economic, and environmental benefits.

KEYWORDS

- **agricultural waste**
- **composting**
- **food supply chain**
- **solid waste**
- **waste recycling**

REFERENCES

Agamuthu, P., (2009). Challenges and opportunities in agro-waste management: An Asian perspective. In: *Inaugural Meeting of First Regional 3R Forum in Asia* (pp. 11, 12).

Amonette, J. E., & Joseph, S., (2009). Characteristics of biochar: Microchemical properties. *Biochar for Environmental Management: Science and Technology*, 33.

ASSOCHAM, (2013). Domestic production of fruits, vegetables to cross 377 MT by 2021: ASSOCHAM (www.netindian.in) (accessed on 3 September 2022).

Balaji, M., & Arshinder, K., (2016). Modeling the causes of food wastage in Indian perishable food supply chain. *Resources, Conservation, and Recycling, 114,* 153–167.

Bellarby, J., Foereid, B., & Hastings, A., (2008). Cool Farming: Climate Impacts of Agriculture and Mitigation Potential. *Environmental Science, 5,* 28–33.

Bhuvaneshwari, S., Hettiarachchi, H., & Meegoda, J. N., (2019). Crop residue burning in India: Policy challenges and potential solutions. *International Journal of Environmental Research and Public Health, 16*(5), 832.

Boateng, A. A., Garcia-Perez, M., Masek, O., Brown, R., & Del, C. B., (2015). Biochar production technology. *Biochar for Environmental Management-Science and Technology*, 63–109.

Chiumenti, R., & Chiumenti, A., (2002). *Technology of Composting*. Agencia regionale per la prevencione e protezione ambientale del Veneto. ISBN 954–8853–59–0, Sofia. (BG).

Dorward, L. J., (2012). Where are the best opportunities for reducing greenhouse gas emissions in the food system (including the food chain)? A comment. *Food Policy, 37*(4), 463–466.

ElMekawy, A., Srikanth, S., Vanbroekhoven, K., De Wever, H., & Pant, D., (2014). Bioelectro-catalytic valorization of dark fermentation effluents by acetate oxidizing bacteria in bioelectrochemical system (BES). *Journal of Power Sources, 262*, 183–191.

Food and Agriculture Organization, (2009). *The State of Food Insecurity in the World*. United Nations Food and Agriculture Organization, Rome.

Gaunt, J., & Cowie, A., (2009). Biochar, greenhouse gas accounting and emissions trading. *Biochar for Environmental Management: Science and Technology*, 317–340.

Gustavsson, J., Cederberg, C., Sonesson, U., Van, O. R., & Meybeck, A., (2011). *Global Food Losses and Food Waste-FAO Report* (pp. 1–37). Food and Agriculture Organization (FAO) of the United Nations.

Hoornweg, D., &Bhada-Tata, P., (2012). What a Waste: A Global Review of Solid Waste Management. *Science Research. 6,* 34–38.

https://www.livescience.com/63559-composting.html (accessed on 30 December 2021).

India, P., (2011). *Census of India 2011 Provisional Population Totals*. New Delhi: Office of the Registrar General and Census Commissioner.

Izaurralde, R. C., Rosenberg, N. J., & Lal, R. (2001). Mitigation of climatic change by soil carbon sequestration: issues of science, monitoring, and degraded lands. *Advances in Agronomy, 70,* 1–75.

Jain, N., Bhatia, A., & Pathak, H., (2014). Emission of air pollutants from crop residue burning in India. *Aerosol and Air Quality Research, 14*(1), 422–430.

Jiang, H., Frie, A. L., Lavi, A., Chen, J. Y., Zhang, H., Bahreini, R., & Lin, Y. H., (2019). Brown carbon formation from nighttime chemistry of unsaturated heterocyclic volatile organic compounds. *Environmental Science & Technology Letters, 6*(3), 184–190.

Joshi, R., & Ahmed, S., (2016). Status and challenges of municipal solid waste management in India: A review. *Cogent Environmental Science, 2*(1), 1139434.

Keener, H. M., Dick, W. A., & Hoitink, H. A., (2000). Composting and beneficial utilization of composted by-product materials. *Land Application of Agricultural, Industrial, and Municipal By-Products, 6,* 315–341.

Korres, N. E., (2020). Utilization and management of agricultural wastes for bioenergy production, weed control, and soil improvement through microbial and technical processes. In: *Environmental Microbiology and Biotechnology* (pp. 143–173). Springer, Singapore.

Lee, P., & Willis, P., (2010). *Waste Arisings in the Supply of Food and Drink to Households in the UK*. Waste & resources action program (WRAP).

Litchfield, J. H., (1987). Microbiological and enzymatic treatments for utilizing agricultural and food processing wastes. *Food Biotechnology, 1*(1), 29–57.

Logan, B. E., & Regan, J. M., (2006). Electricity-producing bacterial communities in microbial fuel cells. *Trends in Microbiology, 14*(12), 512–518.

198 Integrated Waste Management Approaches for Food and Agricultural Byproducts

Lohan, S. K., Jat, H. S., Yadav, A. K., Sidhu, H. S., Jat, M. L., Choudhary, M., & Sharma, P. C., (2018). Burning issues of paddy residue management in north-west states of India. *Renewable and Sustainable Energy Reviews, 81,* 693–706.

Maroušek, J., Strunecký, O., & Stehel, V., (2019). Biochar farming: Defining economically perspective applications. *Clean Technologies and Environmental Policy,* 1–7.

McHenry, M. P., (2009). Agricultural bio-char production, renewable energy generation and farm carbon sequestration in Western Australia: Certainty, uncertainty, and risk. *Agriculture, Ecosystems & Environment, 129*(1–3), 1–7.

McLachlan, M., & Hamann, R., (2011). Theme Issue on Food Security. *Development Southern Africa. 28,* 429–430.

Mena, C., Adenso-Diaz, B., & Yurt, O., (2011). The causes of food waste in the supplier–retailer interface: Evidences from the UK and Spain. *Resources, Conservation, and Recycling, 55*(6), 648–658.

Mittal, S. K., Singh, N., Agarwal, R., Awasthi, A., & Gupta, P. K., (2009). Ambient air quality during wheat and rice crop stubble burning episodes in Patiala. *Atmospheric Environment, 43*(2), 238–244.

Mohan, S. V., Velvizhi, G., Krishna, K. V., & Babu, M. L., (2014). Microbial catalyzed electrochemical systems: A bio-factory with multi-facet applications. *Bioresource Technology, 165,* 355–364.

Obi, F. O., Ugwuishiwu, B. O., & Nwakaire, J. N., (2016). Agricultural waste concept, generation, utilization, and management. *Nigerian Journal of Technology, 35*(4), 957–964.

Pant, D., Singh, A., Van, B. G., Olsen, S. I., Nigam, P. S., Diels, L., & Vanbroekhoven, K., (2012). Bioelectrochemical systems (BES) for sustainable energy production and product recovery from organic wastes and industrial wastewaters. *RSC Advances, 2*(4), 1248–1263.

Papargyropoulou, E., Lozano, R., Steinberger, J. K., Wright, N., & Bin, U. Z., (2014). The food waste hierarchy as a framework for the management of food surplus and food waste. *J. Clean. Prod. 76,* 106–115.

Sahai, S., Sharma, C., Singh, S. K., & Gupta, P. K., (2011). Assessment of trace gases, carbon, and nitrogen emissions from field burning of agricultural residues in India. *Nutrient Cycling in Agroecosystems, 89*(2), 143–157.

Scaglia, B., Tambone, F., Genevini, P. L., & Adani, F., (2000). Respiration index determination: Dynamic and static approaches. *Compost Science & Utilization, 8*(2), 90–98.

Shilev, S., Naydenov, M., Vancheva, V., & Aladjadjiyan, A., (2007). Composting of food and agricultural wastes. In: *Utilization of By-products and Treatment of Waste in the Food Industry* (pp. 283–301). Springer, Boston, MA.

Srinivasarao, C., Venkateswarlu, B., Lal, R., Singh, A. K., & Kundu, S., (2013). Sustainable management of soils of dryland ecosystems of India for enhancing agronomic productivity and sequestering carbon. In: *Advances in Agronomy* (Vol. 121, pp. 253–329). Academic Press.

United, N., (1997). *Glossary of Environment Statistics, Studies in Methods.* United Nations New York, NY.

Vaidyanathan, J., (2018). Impact of Food Wastage on Environment Sustainability. *International Journal of Pharma Research, 11,* 11–17.

Washenfelder, R. A., Attwood, A. R., Brock, C. A., Guo, H., Xu, L., Weber, R. J., & Cohen, R. C., (2015). Biomass burning dominates brown carbon absorption in the rural southeastern United States. *Geophysical Research Letters, 42*(2), 653–664.

Whiting, A., & Azapagic, A., (2014). Life cycle environmental impacts of generating electricity and heat from biogas produced by anaerobic digestion. *Energy, 70*, 181–193.

Zhang, H., Hu, D., Chen, J., Ye, X., Wang, S. X., Hao, J. M., & an, Z., (2011). Particle size distribution and polycyclic aromatic hydrocarbons emissions from agricultural crop residue burning. *Environmental Science & Technology, 45*(13), 5477–5482.

CHAPTER 7

Challenges and Opportunities Associated with Food and Agricultural Waste Management Across the Globe

DRISHTI KADIAN,[1] SYED MANSHA RAFIQ,[2] NIKUNJ SHARMA,[2] SYED INSHA RAFIQ,[1] SYED ANAM UL HAQ,[3] R. M. SHUKLA,[4] FAIZAN MASOUDI,[4] AND INSHA NAZIR[5]

[1] *Department of Dairy Technology, ICAR-National Dairy Research Institute, Karnal – 132001, Haryana, India*

[2] *Department of Food Science and Technology, National Institute of Food Technology Entrepreneurship and Management, Sonipat – 131028, Haryana, India, E-mail: mansharafiq@gmail.com (S. M. Rafiq)*

[3] *Department of Plant Biotechnology, Sher-e-Kashmir University of Agricultural Sciences and Technology, Shalimar, Srinagar, Jammu and Kashmir, India*

[4] *College of Agricultural Engineering Sher-e-Kashmir University of Agricultural Sciences and Technology, Shalimar, Srinagar, Jammu and Kashmir, India*

[5] *Department of Floriculture and Landscaping, Sher-e-Kashmir University of Agricultural Sciences and Technology, Shalimar, Srinagar, Jammu and Kashmir, India*

Integrated Waste Management Approaches for Food and Agricultural Byproducts. Tawheed Amin, PhD, Omar Bashir, Shakeel Ahmad Bhat & Muneeb Ahmad Malik (Eds.)
© 2023 Apple Academic Press, Inc. Co-published with CRC Press (Taylor & Francis)

ABSTRACT

There is significant evidence to substantiate that the world population is growing and will continue to grow at an accelerated pace. Increased food production as a result of this population growth will thus bring the challenge of managing the increase in food waste or loss. A large amount of food that gets thrown away in the open already has severe repercussions on our environment. Food waste has a vast scope to be utilized as a source for energy production and waste valorization. The conversion of food waste is inherently a challenging task for the global food and agriculture service industry. To address this issue, the current chapter explores opportunities for waste management (WM) involving anaerobic digestion (AD), thermal, and thermo-chemical technologies, fermentation, and some novel extraction techniques. In addition, this brief discussion of the concepts can provide a future narrative for more operative management of food waste.

7.1 INTRODUCTION

Food waste can be defined as '*the decrement in amount or quality of food throughout the food supply chain*' (FSC) or in other words, it can be defined as "*portions of food and inedible/ unconsumable parts of food*" (FAO, 2019). Also, these are the by-products generated from food industries, homes, canteens, restaurants, and similar establishments (Banu et al., 2020). In general, 33% of the food produced for human consumption is wasted or discarded throughout the world causing socio-economic problems, environmental hazards and ethical issues (Campos-Vega, 2020). The FSC consists of seven main segments, namely agricultural production, harvest, and post-harvest/slaughter/catch operations, storage, transportation, processing, wholesale or retail, followed by consumption (FAO, 2019). The losses of food occur mainly at the time of manufacturing, post-harvest handling and storage amounting to 54% of the world's food waste while the rest occurs at the marketing, handling, and consumption stages (FAO, 2013). Apart from the wastage of resources, $750 billion annual economic loss is incurred resulting directly from food losses. Excess production of food and dumping injects carbon footprint of around 3.30 billion tons of carbon dioxide (CO_2) equivalent greenhouse gas into the atmosphere (FAO, 2015). Generation of food waste is affected by numerous factors like the behavior of the consumer – the habit of over-buying, over-consumption, over-cooking, unplanned household activities; retailer-throwing expired and unsold food products. Further, there is a low demand for recycled and value-added products from food waste. Food wastage is an unstructured area (Weber and Khademian, 2008), and high waste generation is related to the inefficiency to transform raw material into a useful end product as a whole. Waste generated in the environment from various industries, domestic and commercial, has increased drastically due to the accelerated progress in big-scale industries over the past few decades. Demand for food is increasing with an increase in population which has put immense pressure on urban and rural areas to manage food and agriculture waste (FW). It is becoming more and more imperative to find a definitive solution.

In the recent past, WM was done by dumping the waste in open bins, but now WM has come a long way. The majority share of municipal waste comes from the food sector (Meyer et al., 2013), which they collect and dispose of in a sophisticated way. WM cannot be described in a simple term since it is a long process involving several steps from collection of waste to monitoring it. However, the management of food waste presents an integral opportunity for sustainable development. Given this, the current chapter explores various opportunities and technologies that are used to transform waste food into better quality products.

7.2 WASTE GENERATION STAGES AND SOURCES

There is a continuous waste generated throughout the value chain. The value chain includes all the major steps throughout the movement of food products along the supply chain (Ghamrawy, 2019). The steps include: (i) agricultural production loss, i.e., spilled or damaged agricultural output during harvest, sorting, and handling, (ii) postharvest handling and storage losses including spillage and degradation during handling, storage, and transportation away from the farm, (iii) processing losses including all the losses due to spillage and degradation during processing carried out in industries, including crops sorted out or lost during process interruptions, (iv) distribution losses, i.e., the losses mainly experienced in market system, e.g., in wholesale markets, supermarkets, retailers, and (v) wet markets and consumption waste, i.e., the wastes incurred at the household level, typically due to discards. A large number of processing industries generate wastes. Some of them are discussed.

7.2.1 WASTES FROM DAIRY INDUSTRY

The main wastes generated are whey, dairy sludges, and wastewater (processing, cleaning, and sanitary). They have high nutrient concentration, biological oxygen demand (BOD), chemical oxygen demand (COD) and organic and inorganic contents. Furthermore, they can contain different sterilizing agents and a wide range of acid and alkaline detergents. Pollution due to dairy industry affects the air, soil, and water quality (Ahmad et al., 2019).

7.2.2 WASTES FROM FRUITS AND VEGETABLES INDUSTRY

Fruits and vegetable wastes can be defined as those inedible portions which are discarded during collection, handling, transportation, and processing (Plazzotta et al., 2017). Wastes generated from various fruits and vegetables include peels, stones, seed kernel, skin, pomace, core, outer leaves, rind, etc. (Torres-León et al., 2018). In general, the waste generated from fruits and vegetables are rich in bioactive components, including phenolics, flavonoids, carotenoids which are extracted and utilized employing nanoencapsulation techniques (Saini et al., 2019).

7.2.3 WASTES FROM THE OILSEED INDUSTRY

Oilseeds include olive, cottonseed, peanut, sunflower, rapeseed, palm, soybean, etc. (Fine et al., 2015). These can be processed via mechanical pressing or solvent extraction to produce oil. Either way generates wastes such as stems, pods, leaves, broken grain, dirt, small stones, and extraneous seeds (Galanakis, 2015). The waste materials such as used drilling fluids and drilling cuttings like complex mixtures of clays and chemicals are generated from the oil industry (Lodungi et al., 2016). In the case of palm oil, the main wastes include palm oil mill effluent (POME), empty fruit bunches, fiber, and shell (Sulaiman et al., 2010). The wastes generated at various stages of processing, during solvent extraction, oilseed meal is produced as a by-product, during protein extraction and protein precipitation, solid residue and whey are obtained respectively. Mainly oilcake and hulls are obtained from oils such as sunflower, sesame, pumpkin, rapeseed (Kachel-Jakubowskaa et al., 2016). Sunflower wax is generated from sunflower oilseeds as waste (Redondas et al., 2019). Oilseed meals are obtained from oil processing after pressing or extraction (Oreopoulou and Russ, 2006). The cakes thus left are mostly rich in appreciable amounts of protein, minerals, residual oil and other nutrients which must be recovered and reused in the food industry (Kosseva and Webb, 2020).

7.2.4 WASTES FROM MEAT AND POULTRY INDUSTRY

In general, the wastes are generated in various steps of meat and poultry processing. These include manure and mortalities (during delivery and holding of the livestock for slaughter), blood, and wastewater (stunning and slaughter), hides, feathers, hoofs, heads, and wastewater, dirt (during hide/hair removal/defeathering), offal, viscera, manure, wastewater, paunch content, and liquor (during evisceration/offal removal), flesh, grease, blood, fat, and meat trimmings and wastewater (during trimming, dressing, and carcass washing) and wastewater (during boning and chilling) (Arvanitoyannis and Ladas, 2008).

7.2.5 WASTES FROM CEREALS INDUSTRY

In the case of cereals, the major wastes produced include rice husk, rice straw, wheat straw, oat husk, hulls from milling operations, buckwheat hulls, durum wheat bran. Generally, the cereal wastes are employed for the production of biofuel (biodiesel, biogas, bioethanol, bio ether, biobutanol, syngas, and biohydrogen), bioactive compounds, biodegradable plastics, bio-adsorbent, and stable nanoparticles (Belc et al., 2019).

7.3 OPPORTUNITIES FOR WASTE MANAGEMENT

Food waste has an immense possibility for acting as an energy source. Further, the waste encompasses various attractive components that can be utilized as a substrate or source of nutrients in several value-addition processes. Throughout the literature, current, and traditional practices of waste valorization have been listed and studied, which are landfill, composting, recycling, AD, thermal, and thermo-chemical, fermentation, and extraction. Various techniques owing to their capacity and efficacy to generate valuable and renewable products from waste are discussed here.

7.3.1 ANAEROBIC DIGESTION

Anaerobic digestion (AD) is the most widely used renewable energy technique to treat the waste of rich moisture products (Brennan and Owende, 2010). It is considered a cost-effective renewable energy process to treat the waste of high moisture content (Romero-Güiza et al., 2016). AD is the biological process where a group of specific bacteria break the biodegradable material in the absence of oxygen. Production of methane gas by conversion of food waste by AD process is beneficial for energy generation. The other biological processes, like pyrolysis and incineration, require more energy and harm the atmosphere much more than AD. The AD process involves four steps (Figure 7.1): *hydrolysis, acidogenesis, acetogenesis*, and *methanogenesis* (Kondusamy and Kalamdhad, 2014). In the first stage-hydrolysis, high molecular compounds, which include protein, carbohydrates, and fat gets broken down into simpler complex such as amino acids, glucose, and fatty acids by fermentative bacteria. In the next step, which is acidogenesis, the hydrolyzed simpler compounds are degraded into volatile fatty acids (VFA) by acidogenic bacteria (Kumar et al., 2016). Acidogenesis is followed by the third step, acetogenesis; the VFA gets digested into CO_2, H_2O, and acetate. These are then utilized by *Archaea methanogenics* in the methanogenesis step to produce methane and CO_2 alongside other gases like N_2, O_2 and H_2S which escapes causing environment pollution (Zhu et al., 2009).

The hydrolysis step is generally considered as rate-limiting step since some complex macromolecular degradation takes longer time (Yuan and Zhu, 2016). To overcome this, the pre-treatment of waste is essential to improve biodegradability, which further helps in improving the biogas production. Pre-treatment such as chemical, thermal, non-thermal, physical, and biological, mainly depends on different kinds of substrates to get optimum yield.

Challenges and Opportunities Associated with Food and Agricultural Waste

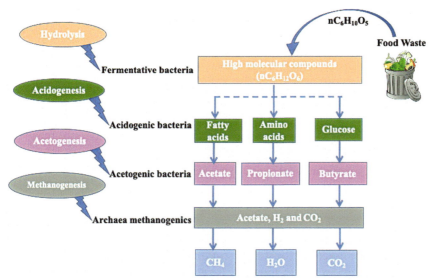

FIGURE 7.1 Anaerobic digestion process.

There are several factors involved which affect the rate of methane production, and to carry out the best metabolic actions is contingent on perfect environmental conditions. Some of the factors to improve biomethane production significantly are listed below:

7.3.1.1 PH

Bacteria's involved in AD needs specific pH to grow efficiently. It has proven that 6.50–7.50 is the optimal pH range to obtain maximum biogas yield. Addition of sodium hydroxide (NaOH) and sodium bicarbonate (NaHCO$_3$) can control the pH.

7.3.1.2 TEMPERATURE

Mesophilic temperature (25–40°C) is necessary to be maintained in an anaerobic digester. AD can be hindered with temperature fluctuation below this range. However, AD is reported to show higher biodegradable potential in thermophilic temperature (50–60°C).

7.3.1.3 FOAMING

Compounds such as proteins, fats, and surfactants are accountable for foaming in the digester (Lindorfer and Demmig, 2016). Production of foaming generally traps biogas and diffuses it into the liquid; this causes blockage in the pipes and also volume reduction in the digester. Kougias et al. (2014) suggested that the anti-foaming agents like long-chain fatty acids, salts, esters, natural oil, and silicone can be used to control foaming. Apart from this enzymatic hydrolysis is also considered a reliable option for foaming problems.

7.3.1.4 NUTRIENTS

Nutrients are essential for smooth running of a digester. Carbon is needed to form energy, whereas nitrogen is required to form protein in the cell. C/N ratio should be between 20–30 to carry out effective AD process, although it also depends on the substrate used (Zhang et al., 2014). Trace elements such as nickel, cobalt, iron, molybdenum, and selenium help in the growth and activity enhancement of methanogenic and fermentative bacteria. According to Xu et al. (2018), FW is not a good source of trace elements, so these can be introduced through sewage sludge or animal manure.

7.3.2 THERMAL AND THERMOCHEMICAL TECHNOLOGY

7.3.2.1 PYROLYSIS AND GASIFICATION

Pyrolysis and gasification both use the action of heat to process waste. These mainly work on carbon-based waste. Pyrolysis converts waste into bio-oil with 95.50% efficiency (Demirbas, 2004) at 400–600°C temperature in the absence of oxygen whereas in gasification the temperature needed is much higher around 800–1500°C in the presence of oxidizing agent (Ruiz et al., 2013). Degradation of waste by pyrolysis results into: (i) syngas ($CO_2 + H_2$) require higher temperature with low heating, (ii) tar or pyrolysis oil from liquid fraction require moderate temperature with high heating rate, and (iii) char – a solid waste residue require low temperature and heating rate. This pyrolytic oil has an enormous probability to be used

Challenges and Opportunities Associated with Food and Agricultural Waste 209

as a base for synthetic crude oil. Along with these products, some light gases like methane (CH_4), carbon monoxide (CO), carbon dioxide (CO_2) and hydrogen (H_2) can also be generated. The three primary conventional pyrolysis approaches are slow, fast, and flash. Nowadays, there is a novel technique called microwave-assisted pyrolysis, which is in demand for food and agriculture waste processing. In order to get more gas yield, lower energy consumption and minimum duration to complete the process, this novel technique is by far the best. This technique can be applied to different waste materials like wheat and corn straw bale, coffee, hulls, sewage sludges and glycerol (Zhao et al., 2011). Gasification process results in hydrogen-rich syngas which can be cast-off as a valuable product into chemicals and fuels (Young, 2010). Reduction of waste material occurs by passing CO_2 or water vapor over charcoal bed to generate combustible gases. Generally, the gasification process converts biomass into gas which is full of carbon dioxide (CO_2), carbon monoxide (CO), hydrogen (H_2), nitrogen (N_2) and methane (CH_4) (McKendry, 2002). Other than this, it also emits char and tar due to partial conversion of biomass. The syngas produced by gasification can be bend into a varied range of products like bio-synthetic gas (bio-SNG), synthetic diesel and hydrogen. Various types of gasifiers used to convert waste are: fixed or moving bed, fluidized bed, plasma reactor, entrained bed. Gasification is more affordable and has the ability to be coupled with a wide range of parameters. This makes it highly advantageous over traditional methods.

7.3.2.2 INCINERATION

Incineration is the most capable technology to convert food and agricultural waste into energy and heat. Waste is subjected to heating in a controlled environment at a temperature range of 870–1,200°C in an enclosed space. The biogas produced during process gets converted into heat energy which can be utilized by industries to produce steam via heat exchangers, and further can also be utilized to generate electrical power (Autret et al., 2007). Incinerator produces several gases like CO_2, nitrogen, carbon monoxide, sulfur dioxide, which can be used efficiently. Incineration shares some impeccable points:

- Reliable technology to get inert ash product from hazardous waste (Arvanitoyannis et al., 2008);

- The waste can be processed in the proximity of waste collection unit itself;
- It has a shorter residence time;
- solid wastes are reduced in volume by up to 80–85%;
- no methane production like it happens in landfill; and
- It is useful in handling any waste like slurries, solid, liquid, sludges, and gases.

Incinerators have two combustion chambers to burn waste material: in the first chamber, combustion is initiated, and waste material gets converted into gas. Combustion gets completed in the second chamber. The ash that remains left after incineration can be treated with a solidification or stabilization method. Various incinerators are available to treat waste material which includes rotary kiln, moving grate, fluidized bed and infrared incinerator. In spite of its advantages, the deposition of hazardous pollutants like dioxins and some suspended particles in the neighboring communities is a major concern.

7.3.2.3 HYDROTHERMAL CARBONIZATION

Hydrothermal carbonization (HTC) is an attractive thermo-chemical conversion process, specifically for waste containing high moisture content between 80–90%. Many researchers are attracted towards HTC since it is a wet process with an exceptional capability to get a higher yield of hydrochar and rich energy reserve (Libra et al., 2011). It requires a temperature range of 200–350°C and duration of 0.20–120 h (Pham et al., 2013). HTC has various benefits over biological processes: reduced carbon footprint, no off-odors production, higher degradation of waste, increased nutrient recovery from nitrogen-rich liquids (which can be further used as fertilizers) and elimination of pathogens and other contaminants due to higher temperature (Libra et al., 2011). By-products from HTC process have end-use applications in lignite coal (Berge et al., 2011) and as an adsorbent for harmful pollutants (Parshetti et al., 2014). To handle the challenging biomass, numerous studies are looking at HTC as a viable option for treatment of sludge (Peng et al., 2016), manure (Cao et al., 2011), starch, and gums (Subagyono et al., 2014).

Challenges and Opportunities Associated with Food and Agricultural Waste   211

7.3.2.4 FERMENTATION

Biological technology is the most promising process for recovery of phosphorus, nitrogen, and carbon from waste. In fermentation, organic compounds get converted into acid, gas, or alcohol in the presence or absence of oxygen (Pleissner and Lin, 2013). Different types of fermentation methods such as continuous, batch, anaerobic, aerobic, and solid-state fermentation (SSF) can be used for the creation of single-cell protein (SCP) from molasses, lemon peel and pulp, whey (Moeini et al., 2004), biogas, biofuel, bio-alcohols, bio-polymers, organic acids, enzymes, antioxidants, pigments, and so on (Kavitha et al., 2017). Growth of microorganism on a moist solid substrate with no available free water is called solid-state fermentation. This approach is promising for obtaining a variety of chemicals and enzymes from food waste (Couto and Sanromán, 2006). It helps in food waste valorization by acid production, e.g., apple pomace can be utilized for the production of citric acid and ethanol (Shojaosodati and Babaeipour, 2002), sugar cane bagasse or whey for lactic acid production, by enzyme production like pectinases from fruit waste, wheat bran, sugar beet pulp (Zheng and Shetty, 2000), other enzyme sources are α-amylase from banana peel, wheat bran, β-glucosidase, xylanase from crop residues like rice/wheat straw (Abdel-Sater and El-Said, 2000); by different flavor production from agricultural solid and coffee husk waste (Christena et al., 2000).

7.3.2.5 EXTRACTION

Food industries have been using extraction techniques from very long time like solvent extraction, steam or hydro-distillation though certain disadvantages are associated with them. The intensifying apprehensions related to environment and health forced the researchers to come up with more green technologies such as supercritical fluid extraction (SFE), ultrasound-assisted extraction (UAE), microwave-assisted solvent extraction (MASE), etc. To extract and separate compounds in SFE, contact of solid raw material with CO_2 (supercritical fluids) is essential, achieved by depressurization. CO_2 is used as a solvent in SFE because of its non-polar and non-toxic nature, high solvating power than hexane, preventive nature against oxidation reaction. The temperature (31.10°C) and pressure (7.38

MPa) of supercritical fluid are higher than the critical state. This technology has its application in extracting carotenoids from apricot pomace, polyphenol extraction from grape and orange pomace (Benelli et al., 2010) and also waste obtained from tomato processing (Shi et al., 2009). UAE uses acoustic energy in the frequency range of above human hearing, i.e., 20 kHz. The basic principle of ultrasound technique is cavitation in liquid: ultrasonic waves create bubbles and then collapse. The collapsing of bubbles creates cavities which generate high-speed waves resulting in a hard impact on a solid surface and damaging it. To improve the extraction of bioactive compounds, UAE can be successfully used in orange and citrus fruit peels (Khan et al., 2010; Sun et al., 2011), straw obtained from wheat (Sun and Tomkinson, 2002), apple pomace (Virot et al., 2010) for the recovery of polyphenols, β-carotene and hemicellulose.

The microwave offers a substantial advantage over the conventional method. The generation of heat during microwave enhances the diffusion power of solvent by breaking hydrogen bonds inside the matrix. Dipole rotation and movement of ions are the fundamental causes of increase in diffusion (Hubaib et al., 2003). Studies in the literature suggested that MASE can be coupled with steam distillation for the extraction of essential oils (Sahraoui et al., 2011). Orange fruit peel, potato waste and peanut skins can be converted into D-limonene and pectin (Pfaltzgraff et al., 2013), phenolic compounds (Wu et al., 2012), respectively.

Apart from these techniques, there are more emerging techniques which can be prominently used for waste valorization such as pressurized liquid extraction, accelerated solvent extraction, solid-phase micro-extraction, pulsed electric field, membrane technology, laser technology and many more.

7.4 CHALLENGES IN WASTE MANAGEMENT

One of the significant issues is waste generated by global food and agricultural sector. Food waste contains high moisture content; processing this waste with incineration, pyrolysis, or other thermal technique can lead to an adverse effect on our environment. Food industries have to work hard in finding ways to valorize waste, which will have a lesser impact on the atmosphere. The new emerging techniques like fermentation, extraction by novel methods have attracted many industries to convert waste into valuable resources. The carbon footprint of the technology can be minimized by having higher productivity, lower production and investment cost and other such strategies. To boost the effectiveness of traditional technologies, they can be coupled with technologies like membrane processing, pulsed electric field and many more. Biofuel is an emerging energy source which can be used to replace other non-renewable resources like coal, tar, uranium, oil, and others. The supercritical extraction technique is becoming popular and could possibly be seen in the near future for the synthesis of esters in biodiesel production. Another possibility for food and agricultural WM is to utilize the end-products formed in other food processes like for energy generation, utilization of acid, incorporation of bio-active components. Of course, the application of these processes could be challenging when applied to food waste, and therefore it is essential to locate similar WM practices pertaining specifically to the food sector.

7.5 CONCLUSIONS

In today's ever-changing world, currently, keen focus should be on finding ways to reduce carbon emission and to prevent further environmental degradation. When considering WM, human activities result in unmanageable complications which require our urgent attention and need to be addressed most productively and efficiently. Our disposal system needs significant modification, and thus, various environmental friendly strategies should be put in place to this end. In order to have the newer technologies to emerge in the near future, one can rely on the vast imagination, ingenuity, and an open hand in the science and engineering segment.

KEYWORDS

- **anaerobic digestion**
- **fermentation**
- **gasification incineration**
- **pyrolysis**
- **waste management**

REFERENCES

Abdel-Sater, M. A., & El-Said, A. H. M., (2001). Xylan-decomposing fungi and xylanolytic activity in agricultural and industrial wastes. *Int. Biodeterioration & Biodegradation, 47*(1), 15–21.

Ahmad, T., Aadil, R. M., Ahmed, H., Rahman, Ur. U., Soares, B. C. V., Souza, S. L. Q., Pimentel, T. C., Scudino, H., et al., (2019). Treatment and utilization of dairy industrial waste: A review. *T. Food Sci Technol., 88*, 361–372.

Arvanitoyannis, I. S., & Ladas, D., (2008). Meat waste treatment methods and potential uses. *International J. Food Sci. Technol., 43*, 543–559.

Autret, E., Berthier, F., Luszezanec, A., & Nicolas, F., (2007). Incineration of municipal and assimilated wastes in France: Assessment of latest energy and material recovery performances. *J. Hazar. Mater., 139*(3), 569–574.

Banu, R., Kumar, G., Gunasekaran, M., & Kavitha, S., (2020). *Food Waste to Valuable Resources: Applications and Management* (1st edn, p. 462). Academic Press.

Belc, N., Mustatea, G., Apostol, L., Iorga, S., Vlăduţ, V. N., & Mosoiu, C., (2019). Cereal supply chain waste in the context of circular economy. *E3S Web of Conferences, 112*, 1–8.

Benelli, P., Riehl, C. A., Smânia, Jr. A., Smânia, E. F., & Ferreira, S. R., (2010). Bioactive extracts of orange (*Citrus sinensis L. Osbeck*) pomace obtained by SFE and low pressure techniques: Mathematical modeling and extract composition. *J. Supercri. Fluids, 55*(1), 132–141.

Berge, N. D., Ro, K. S., Mao, J., Flora, J. R., Chappell, M. A., & Bae, S., (2011). Hydrothermal carbonization of municipal waste streams. *Environ. Sci. Technol., 45*(13), 5696–5703.

Brennan, L., & Owende, P., (2010). Biofuels from microalgae—A review of technologies for production, processing, and extractions of biofuels and co-products. *Renew. Sustain. Energy Reviews, 14*(2), 557–577.

Campos-Vega, R., Oomah, B. D., & Vergara-Castaneda, H. A., (2020). *Food Wastes and By-products, Nutraceutical, and Health Potential* (p. 480). Wiley Publishers.

Cao, X., Ro, K. S., Chappell, M., Li, Y., & Mao, J., (2011). Chemical structures of swine-manure chars produced under different carbonization conditions investigated by advanced solid-state 13C nuclear magnetic resonance (NMR) spectroscopy. *Energy Fuels, 25*(1), 388–397.

Christen, P., Bramorski, A., Revah, S., & Soccol, C. R., (2000). Characterization of volatile compounds produced by Rhizopus strains grown on agro-industrial solid wastes. *Bioresour. Technol., 71*(3), 211–215.

Couto, S. R., & Sanromán, M. A., (2006). Application of solid-state fermentation to food industry—a review. *J. Food Eng., 76*(3), 291–302.

Demirbas, A., (2004). Combustion characteristics of different biomass fuels. *Progress in Energy and Combustion Sci., 30*(2), 219–230.

FAO, (2019). *The State of Food and Agriculture: Moving Forward on Food Loss and Waste Reduction.* Rome. License: CC BY-NC-SA 3.0 IGO.

Fine, F., Lucas, J. L., Chardigny, J. M., RedlingShöfer, B., & Renard, M., (2015). Food losses and waste in the French oil crops sector. *Oilseeds & Fats Crops and Lipids, 22*(3), 1–14.

Challenges and Opportunities Associated with Food and Agricultural Waste   215

Food and Agriculture Organization of the United Nations, (2013). *Food Wastage Footprint: Impacts on Natural Resources: Summary Report*. FAO.

Food and Agriculture Organization of the United Nations, (2015). Global initiative on food loss and waste reduction. *Key Facts on Food Loss and Waste You Should Know* (pp. 1, 2). FAO.

Galanakis, C. M., (2015). *Food Waste Recovery: Processing Technologies and Industrial Techniques* (1st edn., p. 412). Academic Press.

Ghamrawy, M., (2019). *Food Loss and Waste and Learning Guide, 2019* (p. 44). Publisher: Food & Agriculture Org.

Hudaib, M., Gotti, R., Pomponio, R., & Cavrini, V., (2003). Recovery evaluation of lipophilic markers from Echinacea purpurea roots applying microwave-assisted solvent extraction versus conventional methods. *J. Sep. Sci., 26*(1, 2), 97–104.

Kachel-Jakubowskaa, M., Kraszkiewicza, A., & Krajewskab, M., (2016). Possibilities of using waste after pressing oil from oilseeds for energy purposes. *Agri. Eng., 20*(1), 45–54.

Kavitha, S., Banu, J. R., Priya, A. A., & Yeom, I. T., (2017). Liquefaction of food waste and its impacts on anaerobic biodegradability, energy ratio and economic feasibility. *App. Energy, 208*, 228–238.

Khan, M. K., Abert-Vian, M., Fabiano-Tixier, A. S., Dangles, O., & Chemat, F., (2010). Ultrasound-assisted extraction of polyphenols (flavanone glycosides) from orange (*Citrus sinensis L.)* peel. *Food Chem., 119*(2), 851–858.

Kondusamy, D., & Kalamdhad, A. S., (2014). Pre-treatment and anaerobic digestion of food waste for high rate methane production: A review. *J. Env. Chem. Eng., 2*(3), 1821–1830.

Kosseva, M, R., & Webb, C. Preface. In: *Food Industry Wastes: Assessment and Recuperation of Commodities*. Maria R. Kosseva, Colin Webb. 13–14.

Kougias, P. G., Boe, K., Tsapekos, P., & Angelidaki, I., (2014). Foam suppression in overloaded manure-based biogas reactors using antifoaming agents. *Biores. Technol., 153*, 198–205.

Kumar, G., Sivagurunathan, P., Park, J. H., Kim, S. H., Kumar, G., Sivagurunathan, P., & Kim, S. H., (2015). Anaerobic digestion of food waste to methane at various organic loading rates (OLRs) and hydraulic retention times (HRTs): *Thermophilic* vs. mesophilic regimes. *Env. Eng. Res., 21*(1), 69–73.

Libra, J. A., Ro, K. S., Kammann, C., Funke, A., Berge, N. D., Neubauer, Y., Titirici, M. M., et al., (2011). Hydrothermal carbonization of biomass residuals: A comparative review of the chemistry, processes, and applications of wet and dry pyrolysis. *Biofuels., 2*, 71–106.

Lindorfer, H., & Demmig, C., (2016). Foam formation in biogas plants: A survey on causes and control strategies. *Chem. Eng. Technol., 39*(4), 620–626.

Lodungi, J. F., Alfred, D. B., Khirulthzam, A. F. M., Adnan, F. F. R. B., & Tellichandran, S. A., (2016). Review in oil exploration and production waste discharges according to legislative and waste management practices perspective in Malaysia. *Int. J. Waste Resour., 7*(1), 1–8.

McKendry, P., (2002). Energy production from biomass (Part 2): Conversion technologies. *Bioresour. Technol., 83*, 47–54.

Meyer-Kohlstock, D., Hädrich, G., Bidlingmaier, W., & Kraft, E., (2013). The value of composting in Germany: Economy, ecology, and legislation. *Waste Manag., 33*(3), 536–539.

Moeini, H., Nahvi, I., & Tavassoli, M., (2004). Improvement of SCP production and BOD removal of whey with mixed yeast culture. *Electronic J. Biotechnol., 7*(3), 6–7.

Oreopoulou, V., & Russ, W., (2006). *Utilization of By-Products and Treatment of Waste in the Food Indus, 3,* 209–232.

Parshetti, G. K., Chowdhury, S., & Balasubramanian, R., (2014). Hydrothermal conversion of urban food waste to chars for removal of textile dyes from contaminated waters. *Bioresour. Technol., 161,* 310–319.

Peng, C., Zhai, Y., Zhu, Y., Xu, B., Wang, T., Li, C., & Zeng, G., (2016). Production of char from sewage sludge employing hydrothermal carbonization: Char properties, combustion behavior and thermal characteristics. *Fuel, 176,* 110–118.

Pfaltzgraff, L. A., Cooper, E. C., Budarin, V., & Clark, J. H., (2013). Food waste biomass: A resource for high-value chemicals. *Green Chem., 15*(2), 307–314.

Pham, M., Schideman, L., Sharma, B. K., Zhang, Y., & Chen, W. T., (2013). Effects of hydrothermal liquefaction on the fate of bioactive contaminants in manure and algal feedstocks. *Bioresour. Technol., 149,* 126–135.

Plazzotta, S., Manzocco, L., & Nicoli, M. C., (2017). Fruit and vegetable waste management and the challenge of fresh-cut salad. *Trends Food Sci. Technol., 63,* 51–59.

Pleissner, D., & Lin, C. S. K., (2013). Valorization of food waste in biotechnological processes. *Sustainable Chem. Pro., 1*(1), 21.

Redondas, C., Baümler, E., & Carelli, A., (2019). Sunflower wax recovered from oil tank settlings: Revaluation of a waste from the oilseed industry. *J. Sci. Food Agr., 100,* 201–211.

Romero-Güiza, M. S., Vila, J., Mata-Alvarez, J., Chimenos, J. M., & Astals, S., (2016). The role of additives on anaerobic digestion: A review. *Renewable and Sustainable Energy Reviews, 58,* 1486–1499.

Ruiz, J. A., Juárez, M. C., Morales, M. P., Muñoz, P., & Mendívil, M. A., (2013). Biomass gasification for electricity generation: Review of current technology barriers. *Renewable and Sustainable Energy Reviews, 18,* 174–183.

Sahraoui, N., Vian, M. A., El Maataoui, M., Boutekedjiret, C., & Chemat, F., (2011). Valorization of citrus by-products using microwave steam distillation (MSD). *Inn. Food Sci. Emerg. Technol., 12*(2), 163–170.

Saini, A., Panesar, P. S., & Bera, M. B., (2019). Valorization of fruits and vegetables waste through green extraction of bioactive compounds and their nanoemulsions-based delivery system. *Bioresources and Bioprocessing, 6*(26), 1–12.

Shi, J., Yi, C., Xue, S. J., Jiang, Y., Ma, Y., & Li, D., (2009). Effects of modifiers on the profile of lycopene extracted from tomato skins by supercritical CO_2. *J. Food Eng., 93*(4), 431–436.

Shojaosadati, S. A., & Babaeipour, V., (2002). Citric acid production from apple pomace in multi-layer packed bed solid-state bioreactor. *Process Biochemistry, 37*(8), 909–914.

Subagyono, D. J., Marshall, M., Jackson, W. R., & Chaffee, A. L., (2015). Pressurized thermal and hydrothermal decomposition of algae, wood chip residue, and grape marc: A comparative study. *Biomass and Bioenergy, 76,* 141–157.

Sulaiman, F., Abdullah, N., Gerhauser, H., & Shariff, A. A., (2010). Perspective of oil palm and its wastes. *J. Physical Sci., 21*(1), 67–77.

Sun, R. C., & Tomkinson, J., (2002). Characterization of hemicelluloses obtained by classical and ultrasonically assisted extractions from wheat straw. *Carbohydrate Polym., 50*(3), 263–271.

Sun, Y., Liu, D., Chen, J., Ye, X., & Yu, D., (2011). Effects of different factors of ultrasound treatment on the extraction yield of the all-trans-β-carotene from citrus peels. *Ultrasonics Sonochemistry, 18*(1), 243–249.

Torres-León, C., Ramírez-Guzman, N., Londoño-Hernandez, L., Martinez-Medina, G. A., Díaz-Herrera, R., Navarro-Macias, V., Alvarez-Pérez, O. B., et al., (2018). Food waste and byproducts: An opportunity to minimize malnutrition and hunger in developing countries. *Frontiers in Sustainable Food Systems, 2* (Article 52), 1–17.

Virot, M., Tomao, V., Le Bourvellec, C., Renard, C. M., & Chemat, F., (2010). Towards the industrial production of antioxidants from food processing by-products with ultrasound-assisted extraction. *Ultrasonics Sonochemistry, 17*(6), 1066–1074.

Weber, E. P., & Khademian, A. M., (2008). Wicked problems, knowledge challenges, and collaborative capacity builders in network settings. *Public Adm. Rev., 68*(2), 334–349.

Wu, T., Yan, J., Liu, R., Marcone, M. F., Aisa, H. A., & Tsao, R., (2012). Optimization of microwave-assisted extraction of phenolics from potato and its downstream waste using orthogonal array design. *Food Chem., 133*(4), 1292–1298.

Xu, F., Li, Y., Ge, X., Yang, L., & Li, Y., (2018). Anaerobic digestion of food waste: Challenges and opportunities. *Bioresource Technol., 247*, 1047–1058.

Young, G. C., (2010). *Municipal Solid Waste to Energy Conversion Processes: Economic, Technical, and Renewable Comparisons.* John Wiley & Sons.

Yuan, H., & Zhu, N., (2016). Progress in inhibition mechanisms and process control of intermediates and by-products in sewage sludge anaerobic digestion. *Renewable and Sustainable Energy Reviews, 58*, 429–438.

Zhang, C., Su, H., Baeyens, J., & Tan, T., (2014). Reviewing the anaerobic digestion of food waste for biogas production. *Renewable and Sustainable Energy Reviews, 38*, 383–392.

Zhao, X., Zhang, J., Song, Z., Liu, H., Li, L., & Ma, C., (2011). Microwave pyrolysis of straw bale and energy balance analysis. *J. Analyt. App. Pyrolysis, 92*(1), 43–49.

Zheng, Z., & Shetty, K., (2000). Solid-state production of polygalacturonase by *Lentinus edodes* using fruit processing wastes. *Process Biochem., 35*(8), 825–830.

Zhu, B., Gikas, P., Zhang, R., Lord, J., Jenkins, B., & Li, X., (2009). Characteristics and biogas production potential of municipal solid wastes pretreated with a rotary drum reactor. *Bioresource Technol., 100*(3), 1122–1129.

Index

A
α-amylase, 58, 59, 94, 211
α-galactosidase-A, 51
α-linolenic acid (ALA), 39

B
β-carotene, 46, 50, 212
β-fructofuranosidase, 95
β-glucans, 92, 100
β-glucosidase, 211

A
Aspergillus, 97
  awamori, 38, 58–60, 94
    yeast, 95
  flavus, 95
  niger, 58–61, 94–97
    NCIM 616, 59
  oryzae, 38, 58, 94
  tamarii, 94
Acetic acid
  fermentation, 130
  production, 97
Acetobacter, 97
  aceti, 97, 98
Acetobacterium, 97
Acetoclastic, 131
Acetogenesis, 31, 32, 206
Acetogenins, 130
Acetyl residues, 99
Acid
  catalyzed esterification process, 39
  esterification, 37, 38
  forming microorganisms, 121
  hydrolysis, 132
Acidogenesis, 32, 206
Acidolysis, 61
Actic acid, 32, 96
Active
  composting, 182

microorganism, 180, 181
  packaging substances, 47
Adsorbents, 127
Advanced agricultural equipments, 176
Aeration, 57, 88, 131, 180, 182
Aerobic, 33, 88, 97, 123, 131, 135, 178, 180, 211
  biological transformation, 123
  mortification, 33
  process, 88, 180
Affordable clean energy, 3
African nightcrawlers, 183
Agglomerated
  nanoparticles, 101
  particle size, 136
Agrarian
  food business, 41
  nourishment items, 26
  squander, 33
  straw, 25
Agri squander, 35
Agricultural
  activities, 115, 138, 166, 167, 174, 175
  animal production, 146
  by-products, 87, 98, 105, 170, 189
  characterization (agricultural waste), 14
    animal waste, 14
  crop waste, 17
  expertise, 143
  food
    commerce, 41
    processing waste, 15
    waste (AFW), 29, 30, 35, 170, 177, 189, 190
  hazardous toxic agricultural waste, 17
  industry, 195
  landowners, 167
  ministry, 169
  practices, 167
  production, 116, 202, 203

residues, 59, 82, 177
waste, 1–3, 13, 14, 18, 25, 29, 31–33,
60, 62, 81–84, 87–89, 92, 93, 95,
98–100, 105, 113–118, 136–138, 165,
167–169, 173, 175, 177, 183, 184,
189, 195, 196, 201, 203, 209
composition, 14
linked emissions, 189
management system (AWMS), 62,
116–118, 138, 196
Agri-food
commerce, 36
industry, 55
items, 27
squander, 36
wastes, 27, 57, 61
Agri-refuse, 25, 26
Agrobacterium species, 100
Agro-industrial, 27, 62, 86, 96, 101, 104,
105, 115
divisions, 27
processing practices, 27
waste, 39, 101
Agro-residue, 176
Air circulation rate, 123
Alcaligenes latus, 104
Alcohol extraction, 61
Algal overgrowth, 16
Aliphatic-aromatic copolyesters (AAC), 103
Alkaline catalyzed transesterification reaction, 37
Alkenyl cysteine sulfoxides, 51
Alkylresorcinols, 47
Alliumcepa L., 51
Alpha 1–6 bonded D-glucose residues, 94
Alpha-amylase particles, 59
Ambulance
centers, 154
services, 154
Amino acid, 32, 40, 53, 59, 93, 101, 121,
130, 135, 178, 206
Aminolysis, 61
Ammonia (NH_3), 31–33, 121, 123, 157,
158, 176, 177
Ammonium sulfate, 61, 96
Amphibians, 16
Amylases, 57–59, 94, 132

Amylopectin, 58, 94
Amylose, 58, 94
Anaerobic, 4, 9, 21, 28, 30, 31, 84, 97, 98,
106, 121, 127, 129, 135, 158, 168, 181,
202, 207, 211, 213
absorption, 28
bacterial cultures, 135
biological degradation, 31
degradation, 9
digester, 128, 207
digestion (AD), 4, 9, 21, 30–32, 84,
86, 87, 106, 114, 121, 122, 129, 130,
135–137, 158, 168, 202, 206–208,
213
fermentation, 97, 98
microorganisms, 121
Animal
by-product, 6
feed sources, 145
fodder, 126
material, 8
product presence, 8
sector, 10
waste, 14, 18
generation, 15
water, 189
Anthocyanidin, 50
Anthropogenic
provenance, 31
source, 120
Anti-allergenic, 40
Anticoagulant, 54
Anti-inflammatory, 40, 50–52
Anti-microbial, 40, 52
activity, 101
Antimutagenic actions, 51
Anti-osteoporosis, 52
Antioxidant, 40, 41, 43, 46, 47, 50–52, 85,
93, 211
activity, 40
properties, 50
Anti-thrombotic, 40
Apiaceae family, 50
Aquaculture, 15, 25, 114, 138, 152, 156, 157
waste, 15
Aquaphobic characteristics, 47
Aquaponics, 156, 157
Aquatic milieu, 16

Index

Arabinoxylans, 92
Archaea, 130, 206
  methanogenics, 206
Aromatic
  benzene rings, 89
  rings, 40
Arthrobacter paraffinensis, 97
  thermophillus, 97
Artificial
  colorants, 52
  intelligence, 148, 149, 154
Assessing way outs for reducing food
  wastage, 147
  incineration landfills, 159
    recover as biofuel nutrients, 158
  recover energy nutrients, 158
    recover energy, 159
  recycle, 156
    circular food production, 156
    circular food systems, 157
  redistribute, 153
    food safety, 154
    liability concerns, 153
    reprocessing foodstuffs, 155
    resilience novel information tech-
      nology solutions, 154
  source reduction, 148
    artificial intelligence solutions, 148
    intelligent labeling sensors, 151
    intelligent labeling, 150
    reduce food losses, 152
    reduce waste and losses, 148
Associated Chambers of Commerce
  Industry of India (ASSOCHAM), 168
Atherosclerosis, 51
Atmospheric
  CH4 emissions, 31
  heating, 4
  methane emission, 86
Aureobasidium pullulans, 101
Automatic extraction, 59
Avian flu virus, 158

B

Bacillus, 58, 60, 96, 97
  cereus, 155
  licheniformis, 54, 58, 94
  megaterium, 104

subtilis, 58, 59, 94
Bacteria, 33, 54, 57, 59, 88, 104, 121, 128,
  130, 131, 135–137, 155, 178, 181, 206
  contamination, 57, 86
  fermentation, 98, 104
Beta-1,4-glycosidic linkage, 60
Bioactive, 26, 84, 88
  components, 92, 106, 204
  compounds, 41, 89, 92, 183, 205, 212
  substances, 50, 62
Bio-adsorbent, 205
Biobutanol, 159, 205
Biochar, 133, 184
  aids, 184
Bio-collectively mineralizeun processed
  refuse, 34
Biocompatibility, 56
Bioconversion, 42, 55, 122
Biodegradability, 13, 86, 120, 194, 206
  compounds, 179
  material, 206
  organic material, 85
  packaging, 9
  polylactic acid, 98
  thermoplastic polyester, 103
  waste material, 88
Biodiesel, 36–39, 61, 62, 104, 159, 205, 213
  manufacturing (agri-food waste), 36
Biodiversity, 3, 4, 152, 168, 175, 176, 194
Bioelectrochemical systems (BESs), 183
Bioenergy generation, 183
Bioethanol, 60, 159, 205
  industries, 60
Biofertilizer, 60, 128
Bio-filters, 157
Biofuel, 30, 36–40, 138, 145, 159, 205,
  211, 213
  diverted crops, 145
  industries, 93
Biogas, 32, 87, 120, 122, 128, 135, 136,
  158, 168, 177, 205–209, 211
  formation, 135
  production, 30
Biohydrogen, 205
Biological
  activity, 181
  catalysts, 93

catalyzed electrochemical systems, 183
cell destruction, 135
degradation, 121
fermentation process, 98
oxygen demand (BOD), 86, 116, 204
pre-treatments, 122
sludge, 17
technology, 211
Biomagnification, 177
Bio-manures, 123
Biomass, 25, 33, 35, 36, 39, 84, 88, 131, 133, 171, 180, 181, 183, 184, 209, 210
energy, 35
resources, 25
Biomedical applications, 100, 103
Bio-methane, 84, 87, 120
production, 207
Bio-molecules, 56
Biopharmaceutical sector, 51
Bioplastic
industries, 103
production, 103
Bio-plastics, 103, 106
Biopolymers, 98–100, 160
Bioremediation, 138
Bio-synthetic gas (bio-SNG), 209
Biowaste, 29, 158
Black carbon, 184
Blanching (vegetables), 42
Blood
cholesterol, 92
clotting, 54
pudding, 53
sausages, 53
Blue photoluminescence intensity, 105
Bovine spongiform encephalopathy (BSE), 156, 159
Brewery
industry waste, 170
wastewater, 167, 189
Bronchial asthma, 51
Buckwheat hulls, 205
Bulking agent, 171
Buttermilk, 16, 55, 56
By product
dairy industry, 55
fruit vegetable industry, 41
vegetables, 47

C

Chlorella pyrenoidosa, 39
Cupriavidus necator, 104
Caffeic acid, 50, 92
Calcium, 33, 37, 180
carbide (CaC2), 33
Calorific value, 33
Cancerous cells, 46
Candida
guilliermondii, 94
rugosa, 38
Canning industries, 171
Capital
expenditure, 124
footprints, 157
investments, 194
Carbon, 5, 12, 31, 104, 105, 130, 131, 134, 171, 180, 208
dioxide (CO2), 3, 4, 16, 20, 31–33, 35, 85–87, 120–125, 129–131, 133–136, 143, 151, 175, 176, 178, 179, 184, 202, 206, 208, 209, 211
disulfide (CS2), 33
dots, 104
fixation, 160
footprint, 21, 103, 143, 160, 184, 202, 210, 213
monoxide (CO), 16, 17, 85, 124, 133, 176, 209
nitrogen (C/N), 88, 123, 208
sequestration process, 85
Cardiac ailments, 51
Cardiovascular health, 50
Carnaging, 52
Carotenoids, 43, 50, 204, 212
Carrot, 50, 51
enriched products, 51
Caseinates, 55
Cell
membranes, 181
metabolism, 59
Cellulose, 46, 92, 93, 127, 130, 132, 134
Central
Institute of Plastics Engineering Technology (CIPET), 166
Pollution Control Board (CPCB), 166

Index

223

Centrifugation, 37
Cereal
  grains, 171
  processing wastewater, 193
Charas residue, 137
Charcoal, 33, 134, 184, 209
Chemical
  astringency, 96
  cell destruction, 136
  composition, 42
  constituents, 5
  conversion rates, 184
  energy, 183
  fertilizer, 34, 175, 177
    runoff, 167
  free cultivation, 34
  oxygen demand (COD), 204
Chitosan, 54
Chlorosulfonic acid, 39
Chocolate industry wastewater, 193
Cholecalciferol, 54
Citric acid, 42, 96, 97, 211
Civil criminal liability, 153
Clean
  blazing fuel substitution, 36
  sanitation operations, 15
  syngas, 124
Clostridium, 97
Coenzymes, 180
Colloidal particles, 16
Coloring agent, 43
Combustion, 35, 123, 210
  heat, 123
Co-metabolization, 55
Commercial enzymes, 93
Co-morbidities, 155
Complex, 8, 13
  macromolecular degradation, 206
  polysaccharides, 94
Condemned
  carcass fragments, 157
  meat, 53
Contaminated, 185
  surface groundwater, 195
Conventional
  food chains, 154
  food waste management, 30, 120

acetogenesis, 121
acidogenesis, 121
anaerobic digestion, 121
animal feeding, 120
hydrolysis, 121
landfilling, 120
methanogenesis, 121
fossil fuel procedures, 33
methodology, 128
methods (recycling reusing), 85–87,
  128, 212
  animal feeding, 87
  compositing vermi-compositing, 88
  land filling, 86
pyrolysis approaches, 209
strategies, 85
substrate, 101
technologies, 127
Conversion process, 83, 132, 133, 146
Cooling maturation period, 179
Coronary heart disease, 52
Cosmetic
  industry, 54, 84
  pharmaceutical industries, 52
Cost-effective renewable energy process, 206
Crop
  cultivations, 189
  residue, 12, 17, 166, 167, 169, 173, 174,
    176, 177, 189, 190, 211
  waste, 14, 18
Crude agrarian goods, 26
Crustaceans, 16
Cysteine derivatives, 102

D

Dairy
  industry waste, 170
  products, 26
D-arabinofuranoside, 60
Daucus carota, 50
De-crystallize cellulose, 132
Deep learning, 149
Degenerative disease, 50
Demethylating enzymes, 95
Depletion of,
  nutrients, 17
  oxygen, 177

Depressurization, 211
D-galacturonic acid, 50
D-glucose, 94, 99
D-glucuronic acid, 60
Dietary
    fiber, 46, 51, 85, 88, 89, 92, 93, 183
    supplements, 89
Digestible protein source, 54
Dimethyl ether (DME), 39
Direct fermentation, 55
Directorate of Economics Statistics, 173
Disease
    prevention, 153
    transmission, 20
Disfigured
    food wastes, 167
    onion bulbs, 51
Diverse
    agricultural operations, 14
    microbial groups, 179
Docosahexaenoic acid (DHA), 39
Domestic
    consumption, 115
    kitchen refuse, 42
    waste, 6, 12
Drilling
    cuttings, 204
    fluids, 204
Dry rendering, 134
Durum wheat bran, 205
Dynamic market, 156

E

Eatable meat byproducts, 53
Eco-accommodating strategy, 127
Eco-friendly
    biofertilizer, 35
    manner, 192
Ecological
    preferences, 125
    problem, 2, 195
    unbalanced environment, 176
Economic
    development, 18
    generation, 83
Edible
    byproducts, 53
    packaging method, 93

rendering process, 134
vegetables, 152
waste, 7
Eisenia foetida species, 88
Electrical energy, 183
Electrocoagulation, 136
Electron-donating, 183
Elementary phenolic molecule, 40
Encapsulation technology, 93
Energy
    production, 4, 84, 114, 202
    recovery, 33, 84, 85, 93
Environmental
    apprehensions, 1, 3
    conservation body, 84
    degradation, 166, 213
    footprint, 156, 157
        reduction objectives, 148
    guidelines, 28
    hazards, 41, 202
    humane animal production practices, 156
    issues, 167
    menace, 83
    performance, 84
    pollution, 25, 175, 176
    Protection Act, 186
    protocols, 192
    ramifications, 7
    repercussion, 114
Enzymatic, 54, 131, 137, 208
    hydrolysis, 132
    treatment, 54
Eriobotrya japonica Lindley, 94
Eriocitin, 43
Essential
    oils, 95, 183, 212
    soil nutrients, 176
Esterification, 36, 37, 39, 40, 61
    processes, 39
    reaction, 40
European
    Food Safety Authority (EFSA), 155, 159
    Union (EU), 4, 8, 9, 26, 30, 31, 36, 87, 143, 144, 155, 156
Eutrophication, 177
Exopolysaccharides, 98–101

Index 225

Exorbitant, 56
Explicitly planned bioreactors, 121
Extensive agrochemical operations, 167
Extracellular
  enzymes, 180, 181
  medium, 98
Extraction, 211
  techniques, 202, 211
Extraneous seeds, 204

F

Fabrication (bioinoculants), 34
Farm
  plastic wastes, 171
  residues, 167, 169
Fatty acid, 36, 38, 61, 89, 104, 121, 130, 206
  methyl ester, 36
Federation of Indian Chambers of
  Commerce Industry (FICCI), 166
Feed
  conversion, 157
  supplements, 183
Fermentation, 36, 42, 94, 96–98, 100, 101,
  103, 104, 130, 202, 206, 211–213
  apple juice wine lees, 193
  bacteria, 206, 208
  medium, 96
  methods, 211
  microorganisms, 121
  process, 98, 103, 104
Flash pyrolysis, 133
Flavanones, 89
Fluidized bed, 209, 210
Fluorescent nano-materials, 104
Foam-forming capability, 54
Foliar fertilizer, 35
Food
  agricultural
    Organization (FAO), 2, 3, 21, 26, 27,
      46, 82, 113, 114, 142–145, 168,
      171, 172, 175, 194, 202
    scraps, 183
    sector, 165, 189, 212
    waste, 183, 195, 196
  agri-industrial waste, 85
  banks, 153, 154

borne
  disease, 148, 152, 155
  illness, 153
  pathogens, 147
business operators, 148, 156
chain, 4, 113, 143, 145, 149, 153, 156,
  157, 177, 194
depletion, 148
donation, 153, 154, 161
  chains, 154
generated
  metabolites, 151
  services, 27
grade acetic acid, 97
industry, 15, 41, 86, 95, 96, 99–101,
  104, 115, 149, 159, 168, 170, 184,
  193, 194, 202, 205, 211, 212
insecurity, 154, 155
life cycle stages, 4
logistics network, 10
loss (FL), 3, 4, 21, 82, 114, 115, 120,
  122, 124, 128, 142, 145–148, 158,
  172, 175, 202
manufacture, 11
package
  industry, 101
  integrity, 151
pharmaceutical industry, 93
poisoning, 18, 155
poverty, 144
price speculation, 145
processing, 2, 10, 15, 28, 93, 115, 136,
  143, 157, 158, 189, 195
  amenities, 28
  waste, 14, 115, 136
production, 2, 4, 18, 82, 142, 148, 152,
  156–158, 160, 172, 174, 189, 194, 202
  intensification, 152
  system, 156, 157
protection, 144, 151, 158
reprocessing, 156
retail transactions, 148
safety, 18, 142, 145, 147, 149–152,
  154–158, 160
scarcity, 143, 161, 168
security, 82, 83, 142, 144–146, 148–150,
  152–154, 156, 158, 161

service
  establishments, 143
  organizations, 115
shortages, 145, 168
supply chain (FSC), 2, 13, 20, 21, 26,
  113–115, 142–144, 146, 149, 166,
  171, 173, 196, 202
theft, 152
wastage, 1–4, 11, 13, 19–21, 113, 114,
  142, 143, 145–147, 160, 172, 175,
  203
waste (FW), 2–7, 9, 10, 12, 13, 19,
  20, 26–30, 35, 36, 38, 57, 58, 62,
  82–85, 87, 90, 93, 102–106, 113–115,
  117, 119–124, 127, 128, 132, 134,
  141–143, 145, 147–149, 151, 153,
  155–161, 166, 170–172, 175, 178,
  179, 184, 189, 191, 193–196, 202,
  203, 205, 206, 208, 211–213
  compost, 195
  hydrolysate (FWH), 38, 39
  management, 117, 127, 128, 161, 191
  recovery, 161
  reduction techniques, 147
Formic acid, 32
Fractionation processes, 55
Free
  fatty acids (FFA), 36, 37, 39, 40
  radical scavenging, 52
Fructans, 51, 92, 99
Fruit
  peel combinations, 95
  vegetables (F-V), 13, 168
    by-products, 88
    cell walls, 95
    industry, 16
    processing industry, 16
    sector, 10
    waste (FVW), 6, 41, 89, 93, 94, 97,
      101, 122, 123, 126, 127, 170
  washing, 16
Fuelwood, 30

G

Galacturonase, 95
Gallbladder, 53

Gas
  sensors, 151
  stream, 124
  waste, 14
Gasification, 124, 133, 137, 208, 209, 213
  incineration, 213
Generally recognized as safe (GRAS), 56,
  98, 103
Genetic records, 149
Global
  evaluation, 39
  food
    crisis, 128
    wastage, 172
    waste, 144, 195
  greenhouse emissions, 145
  podium, 21
  population, 152
  scenario (waste generation), 1
  staple commodities, 159
  warming, 3, 4, 28, 86, 114, 122, 175,
    176
  waste generation, 3
Glucoamylase (GA), 3, 43, 58, 94
Gluconobacter sps., 42
Glucuronic acid, 99
Glycerophospholipids, 61
Glycine, 53
Granular charcoal, 33
Green
  environment, 82
  technology, 193
Greenhouse
  emission, 1, 3, 122
  gases (GHG), 4, 15, 16, 27, 34, 85, 125,
    142, 174–176, 184
    emission, 82, 83, 88, 128, 168, 184,
      189, 192
Groundwater
  aquifers, 4
  bodies, 4

H

Halotolerant, 59
Hazardous
  pollutants, 210
  pollution, 177

Index

Heart
  ailments, 47
  defensive, 40
Heat
  exchangers, 209
  production, 30
  retention, 185
Hemicellulose, 46, 92, 93, 212
Herbicides, 14
Hesperidin, 43
Heteropolysaccharide, 50, 56, 99
Heterotrophic
  microalgae, 39
  microorganisms, 103
High density polyethylene (HDPE), 103
High-energy waste, 123
High-tech equipment, 158
Homogeneous
  alkaline catalyst, 37
  heterogeneous catalyst, 131
Homopolysaccharides, 99
Horticultural, 27
  products, 89
  waste, 99
Human
  consumption, 7, 18, 82, 98, 105, 144,
    153, 171, 202
  deoxyribonuclease I, 51
  health, 17, 20, 21, 85, 89, 101, 176, 177
Humanity, 2, 114
Humic substances, 85, 123
Humidity, 88, 178, 181
Hydrochar, 210
Hydrocolloids, 50
Hydro-distillation, 211
Hydrogen (H_2), 5, 12, 32, 84, 85, 121–124,
  130, 131, 133, 137, 184, 208, 209, 212
  sulfide, 121
Hydrogenotrophic, 131
Hydrolysis, 31, 32, 37, 58, 59, 61, 94, 121,
  122, 129, 130, 137, 206, 208
  aerobic degradation, 31
  fermentation, 31
  reaction, 37
Hydrolyzation (triglycerides), 37
Hydrophobic compounds, 131
Hydrothermal, 131
  carbonization (HTC), 210

Hydroxycinnamic acids, 50
Hydroxyl groups, 40
Hygiene standards, 155

I

Identical monomeric units, 99
Immune-stimulator, 100
Improper waste disposal, 192
Incineration, 35, 84, 85, 114, 123, 124,
  159–161, 166, 177, 195, 206, 209, 210,
  212
Incinerator, 209
Industrial
  sectors, 82, 84, 85
  waste
    sullage run-off, 117
    water, 16
Industrialization, 16, 18
Inedible
  rendering process, 134
  waste, 7
Inert vitrified glass, 133
Inexhaustible sources, 36
Inexpensive inorganic fertilizers, 18
Infrared incinerator, 210
Infrastructure industry innovation, 3
In-line pipe washing, 15
Inorganic
  fraction, 124
  impurities, 87
Insecticides, 14, 17, 18, 115
Insoluble
  dietary fibers, 46
  fibers, 92
  substrates, 57, 94
Installation
  components, 190
  costs, 118
  storage facility, 191
Intense pollution difficulties, 171
Intergovernmental panel on climate change
  (IPCC), 189
International
  collaboration, 195
  commitments, 4
  monetary fund, 144

228 Index

Intestinal microflora, 47, 92
Intracellular enzymes, 180
Invertase, 57, 95, 96
Invertebrates, 178
Irrigation water, 152
Isoflavones, 89
Isorhamnetin
  3,4-diglucoside, 52
  4-glucoside, 52

J

Jatropha, 40, 60

K

Kaempferol, 52
Karanja, 40
Kidney, 26, 53
Korean potato peel, 92

L

Labor requirements, 118
Lack of,
  communication, 11
  planning, 11
Lactic acid, 32, 98, 103, 211
  polymerization reaction, 98
Lactobacillus, 96, 97
  casei, 98
Lactose, 55, 95
Land
  filling, 31
  spreading, 9
Landfill space energy, 192
L-arabinose, 50
Laser technology, 212
Leachate
  formation, 87, 88
  production, 122
Lecithin, 17
Light absorption, 176
Lignocellulosic
  substrate, 57
  waste, 104
Limited economic resources, 154
Linear aliphatic polyester, 103
Lipase, 61
  enzyme, 61

Lipid commodity, 134
Listeria monocytogenes, 154
Livestock, 2, 14, 25, 30, 62, 87, 114, 138,
  153, 156, 172, 173, 185, 205
  excreta, 25
Local ambient water quality standards, 31
Long-chain fatty acid, 32, 208
Low calorie ingredient, 100, 101
Low density polyethylene (LDPE), 103
Low-cost organic adsorbent, 34
Low-income
  countries, 19, 20
  populations, 144
  urban communities, 145
L-rhamnose, 50
Lupeol, 46

M

Maceration process, 92
Malnutrition, 144, 148, 152, 154, 169
  threats, 144
Maltose, 58, 94
Maltotriose, 58
Malus domestica, 43
Management costs, 191
Mangifera indica L. *anacardiacea*, 46
Mannose, 99
Maturation, 178
Meat, 8, 13, 15, 18, 26, 116
  byproducts, 8
  industry, 15, 52, 53, 92
  packing plants, 134
  products, 8, 54, 100, 153
  protein requirements, 30
  trimmings, 205
Mechanical
  damage, 10
  grinding, 122
  peeling, 51
  resistance, 103
Medical application, 104
Medicinal industries, 47
Membrane
  processing, 213
  technology, 212
Mesophilic temperature, 207

Index

Metabolic waste, 15
Metal
  nanoparticles, 101
  oxide nanoparticles, 101
Metalloids, 126
Metallo-organic molecules, 93
Metallurgical industry, 33
Metamorphose, 33, 35, 38, 57
Methane (CH_4), 4, 14, 17, 28, 31, 32, 54,
  83, 85–87, 114, 120–122, 129–131, 134,
  137, 175, 176, 184, 206, 207, 209, 210
  gas, 86, 137, 206
    production, 137
Methanogenesis, 31, 32, 130, 206
Methanol, 33, 39
Method
  transformation, 125
  treating wastes, 177
    animal feed, 184
    biochar production, 184
    bioelectrochemical systems (BESS),
      183
    composting parameters, 179
    composting, 177
    methods of composting, 182
    value-added processing, 183
  wastewater operations, 28
Methogenesis, 87
Methoxy residues, 95
Microalgae, 36, 38
Microbial
  activity, 28, 180–182
  decomposition, 180
  fermentation, 137
  fuel cells, 183, 194
  growth processes, 181
  pollution, 126
  processing chemicals, 96
  spoilage, 46
  stability, 181
Micronutrients, 144
Microorganism development hazard, 126
Microwave-assisted
  pyrolysis, 209
  solvent extraction (MASE), 211, 212
Milk industry, 15
Milling operations, 205

Million
  metric tons (MMT), 39, 168
  tons (Mt), 2, 19, 31, 41, 86, 115, 119,
    143, 156, 169, 173, 174, 189
Mini-carrots, 50
Ministry of New and Renewable Energy
  (MNRE), 169
Mixed
  food waste, 6
  method, 182
  product, 8
Modern incineration plants, 123
Modernization, 26, 61
Molasses, 17, 100, 167, 189, 211
  wastewater, 170, 189
Molybdenum, 208
Momentous food wastage, 2
Monoalkyl ester, 36
Motor-generators, 120
Multiple food operations, 171
Municipal
  corporations, 120
  waste, 122, 167, 203
Muscular degeneration, 51

N

Nanoencapsulation techniques, 204
Nanoparticle, 102, 106
  synthesis, 102
Nano-range particles, 102
Naringin, 43
Narirutin, 43
National
  Engineering Environmental Research
    Institute (NEERI), 166
  policy for management of crop residues
    (NPMCR), 173, 189, 190
Natural
  anaerobic eruption, 120
  colorants, 85
  compost, 127
  ecosystem, 94
  preservative, 97
  products-meat, 26
  resource, 3, 114, 192
    degradation, 83
    expenditure, 85

Neural backpropagation networks, 149
Neurosporacrassa CFR 308, 59
Neutralization process, 17
Nickel, 208
Nitrogen, 28, 42, 88, 99, 123, 124, 176, 177, 180, 182, 184, 208–211
  phosphorus-potassium (NPK), 88
Noble moral dimension, 83
Non-biodegradable packaging, 9
Non-catering waste, 9
Non-profit group, 153
Non-renewable resources, 213
Normal ecological standards, 31
Nourishment
  added substances, 36
  squander, 27
Novel
  commodities, 28
  implementation, 149
  methodology, 128
  technique (food waste management), 128
    acetogenesis, 130
    acidogenesis, 130
    aerobic digestion, 131
    anaerobic digestion, 128
    biological liquefaction, 132
    bioremediation, 135
    conditioning, 135
    electrocoagulation, 136
    enzymatic hydrolysis, 129
    enzymatic liquefaction, 132
    gasification, 133
    hydrolysis, 132
    hydrothermal liquefaction (HTL), 131
    liquefaction, 131
    methanogenesis, 130
    pyrolysis, 133
    rendering, 134
Nucleic acids, 121, 180
Nutraceuticals, 89
Nutrient, 208
  availability, 34
  balance, 16
  composition, 5
  repossession, 160
Nutritional
  advantages, 87
  edible food, 147

functional characteristics, 55
status, 154
value, 26, 50

O

Odor emissions, 88
Off-odors generation, 126
Oil
  effluent waste, 170
  industry, 16
  producing microorganism, 38
    culture, 38
  refining, 61
  seeds, 204
Oligosaccharides, 94
On-farm crop residue burning, 189
Open
  dumping, 20
  packaging, 6
Operation, 119
  parameters, 56
Organ transplant individuals, 151
Organic
  acids, 96, 97, 121, 130, 137, 151, 178, 181, 211
  contaminants, 158
  crop residue, 6
  fraction (waste), 88
  inorganic contents, 204
  materials, 14
  matter, 31, 35, 42, 85, 86, 88, 121–124, 131–133, 135, 179–182
    degradation, 181
    waste, 133
  solvent, 46
  substances, 32, 33, 85, 121, 179
  translation, 179
Organization for Economic Cooperation Development (OECD), 14
Organoleptic, 56
Organosulfur constituents, 51
Osteoporosis neurodegenerative disorders, 52
Overland-water resources, 195
Oxidation, 35, 47, 105, 135, 184, 211
  agents, 99
  fermentation, 97

Index 231

product, 52
stress-induced damage, 52
Oxygen impermeable films, 101

P

Packaging, 6, 9, 12, 13
bio-degradability, 9
Palm oil
industry, 38
mill effluent (POME), 189, 204
Partial
conversion of biomass, 209
feed substitute, 87
Particulate matter (PM), 176
Pastoral production, 153
Pectic substances, 95
Pectinolytic enzymes, 57
Peroxidases, 57
Penicillium sp, 60, 97
species, 60
Per capita solid waste generation, 18
Perilous lethal agrarian squander, 27
Permeable vegetable structure, 127
Peroxidases, 57
Pest infestation, 182
Pesticides, 14, 17, 18, 115, 117, 167, 175, 185, 186
Petrochemical plastic waste, 104
Petroleum
refineries, 33
wastes, 185
Pharmaceutical, 56, 101
compounds, 33
industries, 54, 96, 97, 100
Phenolic
acids, 43, 89
complexes, 40
compounds (agri-food byproducts), 40, 46, 47, 60, 88, 89, 92, 212
substances, 50, 60
Phenylalanine, 40
Phloridzin, 46
Phosphodiester bond, 61
Phospholipid, 56, 61
production, 61
Phosphorus, 42, 177, 180, 211
Photoluminescence lifetime, 105

Photosynthesis rate, 177
Phytochemical, 52, 84, 89, 183
compounds, 88
Phytoestrogens, 89
Pineapple, 95
cannery, 42
Plant
food, 89
waste, 156
materials, 8
non-digestible polysaccharides, 92
pathogens, 182
Plantation, 25
Plasma assisted gasification (PSG), 133
Plastic, 103, 123, 185, 205
cans, 186
wraps, 185
Policy interventions, 194
Political
harmony, 144
public awareness, 120
Poly ε-caprolactone (PCL), 103
Poly(hydroxyalkanoates) (PHAs), 103, 104
Polybutylene
adipate-terephthalate (PBAT), 103
succinate (PBS), 103
adipate (PBSA), 103
Polycyclic carcinogenic aromatic hydro-carbons, 17
Polyethylene terephthalate (PET), 103, 104
Polyglycolic acid (PGA), 103
Polyhydroxy
alkanoates (PHA), 103
butyrate, 55
Polylactic acid (PLA), 103
Polymer
materials, 101
organic material, 32
Polymethylene adipate-terephthalate (PTMAT), 103
Polypeptide chains, 59
Polyphenol, 89
ascorbic acid, 46
oxidases, 57
Polysaccharides, 60, 94, 99, 101, 121
Polystyrene (PS), 103
Poly-unsaturated fatty acids (PUFAs), 39

Pomace, 16, 43, 46, 50, 60, 61, 84, 89, 93, 95, 97, 104, 204, 211, 212
Post
 agricultural wastage, 144
 harvest, 4, 13, 114, 203
  handling storage, 10
  slaughter-catch operations, 202
 utilization, 27
Potential
 bioceutical attributes, 47
 resource, 25
Poultry
 droppings, 84
 processing, 26, 54, 205
Power heat incineration, 159
Prebiotic characteristics, 51
Pre-consumer waste, 189
Pressurized liquid extraction, 212
Pre-steam process, 59
Pre-treatment
 activities, 118
 practices, 37
Pre-utilization process (wastes), 27
Proanthocyanidins, 40, 47
Probiotic organisms, 92
Product
 development, 93
 yield transesterification method, 38
Production, 10, 13, 30, 117, 190, 206, 208
 biodiesel, 36, 159
 cost, 2
Propionic, 103, 121
Proteases, 57, 59, 60
Protein
 extraction, 205
 precipitation, 205
Proteolysis, 59
Protocatechuic acids, 52
Protozoa, 33, 88
Pseudomonas
 hydrogenovora, 55
 oleovorans, 104
Pseudo-plasticity, 99
Pullulan, 100, 101
Pulsed electric field, 212, 213
Pyramid hierarchy, 125

Pyrolysis, 33, 62, 124, 138, 184, 206, 208, 212, 213
 oils, 33, 208
Pyruvic acid, 32
Pyruvil, 99

Q

Quality standards, 11
Quercetin, 43, 52
 3,4-diglucoside, 52
 3-glucoside, 52
 4-glucoside, 52

R

Radiant thermal energy, 35
Radiofrequency identification detectors (RFID), 151, 152
Raffinose, 42
Rainwater infiltration, 87
Rate of
 accumulation, 86
 decomposition, 180, 182, 186
Raw agricultural materials, 138
Ready-to-eat meals, 8
Re-composition products, 184
Recreational furniture, 185
Refuse
 management, 54
 shrinkage, 30
 water process plants, 32
Regulatory
 dispensing regulations, 31
 economic considerations, 143
Remote sensing data, 149
Renewable
 agricultural sources, 103
 energy
  generation, 183
  technique, 206
Residual pesticides, 185
Resilient grocery chains, 145
Resource footprints, 152, 156
 resilience, 152
Retail systems, 82
Retention fertilizer value, 137
Reusable resources, 167

Index

Rhizopus oryzae, 38, 94
R-hydroxyalkanoic acids, 103

S

Schizochytrium mangrovei, 39
Saccharification, 132
Saccharomyces cerevisiae, 58, 97
Safe
  end products, 135
  to-eat substances, 171
Salvaging, 29
Sanitization, 15, 16
Saponification, 38
  reaction, 36
Sapota, 95
Seasonal
  autoregressive models, 149
  commodity, 46
  components, 157
  variations, 176
Selective porous membranes, 55
Semi-finished goods, 143
Sensor systems, 151
Sewage water treatments, 34
Short-chain
  aliphatic carboxylic acid (SCACA), 42
  fatty acid, 32
Silica-gel matrix, 38
Silicon carbide (SiC), 33
Simarouba, 40
Simple sugars, 132
Single
  cell protein, 43, 183, 211
  product, 8
  stage digesters, 122
Skin cancers protection, 47
Slow
  composting, 182
  pyrolysis, 134
Smart
  intensification, 152
  packaging, 151
Social
  contemplation, 7
  development, 144
  dimensions resilience, 152
  instability, 144

Societal justifications, 147
Socio-economic
  groups, 154
  problems, 202
  safety nets, 145
  stress, 144
Sodium
  bicarbonate, 207
  cyanide (NaCn), 33
  hydroxide, 207
  phosphate, 61
Soil
  alteration, 124, 184
  composure, 183
  fertilization, 122
  productiveness, 176
Solid
  organic substrates, 122
  phase micro-extraction, 212
  recalcitrant material, 123
  state fermentation (SSF), 42, 57, 59,
    93–95, 211
    bioreactor design, 57
  waste, 2, 14–18, 20, 28, 60, 85, 86, 116,
    119, 124, 131, 133, 135, 166, 189,
    196, 208, 210
    treatment, 133
Soluble
  complex organic wastes, 183
  dietary fiber, 50
  sugars, 16
Solvent extraction, 204, 211, 212
Stabilization
  method, 210
  organic refuse, 34
Stable
  functional yogurt, 100
  nanoparticles, 205
Staphylococcus aureus, 155
Steam distillation, 212
Storage, 118, 190
  trucks, 15
Submerged fermentation (SmF), 57, 93
Sulfuric acid, 39
Supercritical
  extraction technique, 213
  fluid extraction (SFE), 211

Surface water bodies, 4
Surplus waste, 189
Suspended solids, 28, 115
Sustainability, 1, 7, 21, 82, 83, 85, 127,
    142, 145–148, 152, 153, 156, 158–161,
    193
    cities communities, 3
    development, 3, 83, 85, 142, 153, 203
        goal (SDG), 3, 4, 21, 83
    water management, 3
Swedish Institute for Food Biotechnology,
    27
Syngas, 124, 205, 208, 209
    polishers, 133
Synthetic natural gas production, 33

T

Tannase, 57, 93, 96
    production, 96
Tannic acid, 93, 96
Tannin, 46, 47, 61, 89, 96, 101
    acyl hydrolase, 96
Targeted delivery (medicines), 101
Technological protocols, 106
Temperature, 3, 5, 11, 12, 15, 33, 57, 59,
    88, 94, 96, 99, 100, 121, 124, 130–135,
    137, 143, 148, 149, 151, 158, 176,
    178–180, 182, 207–211
    fluctuation, 207
    storage, 5
Terpenes, 89
Terrestrial ecosystems, 3
Thermal
    decomposition, 124
    depolymerization process, 131
    treatments, 35, 87, 123, 124
Thermo-chemical
    conversion process, 210
    technologies, 202
Thermo-irreversible hydrogels, 100
Thermomyces lanuginosus, 94
Thermophiles, 180
    stage, 180
    temperature, 207
Thermos-solid, 59
Time-temperature indicators (TTI), 151

Titratable acidity, 98
Toxic
    agricultural waste majority, 2
    waste, 62
Traditional
    cereal meat, 146
    processing protocols, 101
    quantum dots, 105
    statistical models, 148
    technologies, 213
Transesterification, 36, 37
    method, 38, 39
    reaction, 40
Transportation, 10, 11, 14, 82, 83, 86, 120,
    142, 154, 157, 171, 185, 191, 194, 196,
    202–204
Treatment, 9, 13, 118, 191
    sludge, 210
Tricarboxylic acid cycle, 97
Trichoderma harzianum 1073 D3, 61
Trichoderma viride, 60, 61
Triglyceride, 36
Tyrosine, 40, 53

U

Ultrasonic waves, 212
Ultrasound
    assisted extraction (UAE), 92, 211, 212
    degradation, 135
    technique, 212
Ultraviolet (UV), 105
Umbelliferae, 50
Underground table water, 87
United
    Kingdom (UK), 6, 9, 11, 36
    Nations framework convention on
        climate change (UNFCC), 3, 4
    States
        Department of Agriculture (USDA),
            116, 177
        Environmental Protection Agency
            (USEPA), 13, 31
Urbanization, 18, 26, 61, 145
Utilization, 36, 98, 191
    agriculture waste, 87

Index

235

V

Valeric acids, 121
Valorization
  approaches, 85
  methods, 85
Value-added
  chemicals, 85
  product, 40, 183, 195, 203
Vasodilatory impacts, 40
Vegetable
  fruit waste, 184
  washing water, 42
  waste, 189
Vermicomposting, 34, 88
  practice, 34
  treatment, 34
Vermiculture, 138
Vermin-composting, 34, 123, 183
Virgin vegetable oil biodiesel, 37
Volatile
  fatty acid (VFA), 32, 42, 121, 206
  organic compound (VOC), 17, 176

W

Waste
  animal tissue, 134
  collection, 87, 138, 190, 210
  generation, 2, 19
  hierarchy, 84, 191, 192
  management (WM), 1, 20, 25, 46, 62,
    81, 83, 84, 88, 98, 106, 113, 116, 119,
    120, 124, 128, 129, 136, 141, 142,
    160, 165, 166, 168, 191, 193–195,
    201–203, 213
    hierarchy, 83
  minimization, 2, 125
  organic matter, 122
  organization, 29
  recycling, 84, 196
  remediation, 183

to-wealth strategies, 85
treatments, 20
valorization, 35, 202, 206, 211, 212
water, 15, 16, 55, 115
  generation, 15
  purification, 15
  stream, 136
  treatment, 183
  zappers, 133
Water
  footprints, 157
  holding capacity, 100
  lipid expression, 61
  managing
    schemes, 176
    systems, 167
  Resource
    Action Program (WRAP), 6, 185
    recovery facility (WRRF), 121
  retention capacity, 184
Wet
  markets consumption waste, 203
  rendering, 134

X

Xanthan gum, 99
Xanthomonas campestris, 56, 99, 100
Xylan, 60
  spine, 60
Xylanases, 57, 58, 60
Xylose-residue, 60

Y

Yarrowia lipolytica, 97

Z

Zinc oxide, 101, 102
Zoonotic infections, 156